About the author

Chris Goodall is a world-leading expert on new energy technologies. His previous book for Profile, *Ten Technologies to Fix Energy and Climate*, was one of the *Financial Times'* Books of the Year. *How to Live a Low Carbon Life* won the 2007 Clarion award for non-fiction. He publishes Carbon Commentary, a website and newsletter on energy efficiency and advances in renewables. He is also an investor in young companies in the low carbon world.

www.carboncommentary.com

The Switch

First published in the UK in 2016 by
Profile Books
3 Holford Yard
Bevin Way
London WC1X 9HD

Typeset in Bembo, THE Sans and Crete Round to a design by Henry Iles.

10 9 8 7 6 5 4 3

A CIP record for this book is available from the British Library.

288pp

ISBN 978 1 781256350
eISBN 978 1 782832485

Printed in the UK by CPI Group (UK) Ltd, Croydon CR0 4YY,
on Forest Stewardship Council (mixed sources) certified paper.

Mixed Sources
Product group from well-managed
forests and other controlled sources
www.fsc.org   Cert no. TF-COC-002227
© 1996 Forest Stewardship Council
FSC

The Switch

Chris Goodall

PROFILE BOOKS

Contents

Introduction
Cheap solar changes everything

'A cheap, clean source of energy would change everything,' declared Bill Gates in February 2016. The annual letter from his charitable foundation focused on the need for a revolution that gives abundant power to all without further disruption of the climate. Gates is in no doubt of the importance of the challenge, particularly for those currently without access to electricity, and optimistic that the solution will be found. 'Within the next fifteen years – and especially if young people get involved – I expect the world will discover a clean energy breakthrough that will save our planet and power our world.'

It won't take fifteen years. The argument of this book is that the breakthrough that Gates anticipates has already occurred. The world now has a set of technologies that will offer cheap and clean power to all, twenty-four hours a day, twelve months a year. From the *favelas* of Latin America and villages of India to the cities of Europe, solar power offers electricity that is now competitive with all other energy sources. And it is becoming cheaper each month through predictable technological changes. Researching this book I talked to no one who thought these improvements would slow down, let alone stop.

1

The speed of change is remarkable, given solar's somewhat slow beginnings. The basic technology has been around for more than half a century. Vanguard 1 – the fourth satellite ever to be launched – carried six solar panels into orbit in 1958. This was the first time the world had ever used photovoltaics (PV) for a real purpose. The little rectangles of silicon produced a maximum half a watt, a tiny charge that nevertheless enabled the satellite to send data about the composition of the atmosphere back to Earth for the next six years. These primitive panels cost many thousands of dollars per watt. However, by the mid 1970s, the figure had fallen to $100 a watt. Now the cost is about 50 cents and the decline continues. We can generate electricity from the sun's rays at costs which seemed

Solar panels with a purpose: GPS pioneer Roger L Easton (left) inspects the Vanguard 1 satellite with its six silicon solar panels.

utterly unimaginable even a few decades ago. Unimaginable even now, it seems, given the conservatism of almost every estimate that appears in financial reports or the press. In my research I found only one historic forecast of future declines in PV costs that had overestimated the speed of the change.

Nevertheless, perhaps we should have anticipated the rapid decline in solar costs in this decade. Peter Eisenberger, now a professor at Columbia University in New York, co-wrote a paper for his employers at the oil company Exxon in 1989 that predicted solar energy (probably to heat water rather than generate photovoltaic electricity) would become cost competitive with fossil fuels by 2012 or 2013. 'Technology evolves in a surprisingly regular way,' he recently told *Bloomberg News* when explaining his prediction. Much of the first chapter of this book is about why we can be confident that PV cost reductions will continue and make THE SWITCH inevitable. Every time the world's accumulated total of solar panels has doubled, the cost has reliably declined by about twenty per cent.

In the sunnier parts of the world solar photovoltaics already offer electricity at lower total cost than other forms of power. As I write, a Californian utility has just announced another record US low for the price paid for a megawatt hour of electricity from PV farms. Subsidised by a tax break, the figure of about $37 per megawatt hour is said to be the lowest long-term purchase agreement for US electricity. Recent published prices for solar electricity in places as diverse as Chile, India and Brazil show equally striking declines – and, crucially, to levels lower than conventional electricity generators.

Even in the gloomy countries of northern Europe, light from the sun will soon provide electricity at prices lower than fossil fuel alternatives. Fraunhofer, a sober German research institute, sees the cost of solar power from solar parks in south Germany falling as low as 4 euro cents (about 3.2 pence) per kilowatt hour by 2020, a cost that beats any competing source of electricity from a new power plant. In Britain the dramatic fall in the price of solar panels has already pushed PV almost to cost parity with planned gas-fired power stations. Power from wind turbines in the best locations on Atlantic coasts is currently cheaper but even this will be beaten as PV continues its relentless downward march.

The falling cost of a PV panel is a major part of this change: both in its production costs and financing. Solar farms can now be financed at far lower rates of interest than any other source of electricity. Because PV is so utterly reliable and almost maintenance-free, it is a perfect investment for pension funds seeking consistent yearly returns for the thirty-five years of a panel's life. As other investment opportunities around the world have dried up in recent years and interest rates have fallen to unprecedentedly low levels, solar farms have become able to raise money at cheaper and cheaper rates. This is surprisingly important. Cutting the annual interest charge from 8 per cent to 5 per cent reduces the underlying cost of producing electricity from solar installations by almost a third.

But, as solar sceptics never tire of saying, cheap daytime-only power isn't enough. We need the capacity to store solar-generated power for use when the sun isn't shining, or other renewable sources that complement the sun's power. In places like the UK, this means using wind and plant material for substantial amounts of energy. Importantly, we'll also need huge amounts of electricity storage, both for overnight and for throughout the dark seasons of the year.

We also need to be able to manage our electricity grids so that they can cope with the impact of unpredictable supplies during the daylight hours. This means finding ways of adjusting electricity demand to match the available supply, not the other way round.

The good news is that this is all well within our reach now, and at costs that seem reasonable today and which will get less expensive every year. Tesla, for example, the electric car and battery company, is offering energy storage at prices that are falling at least as fast as the costs of PV. Rapid advances in complementary technologies such as wind, hydrogen production and anaerobic digestion mean that the world will not lack power when the sun is down. The second part of this book focuses on these areas of storage and non-solar renewables.

In considering the strengths and vulnerability of solar, it is important to keep in mind where the world's population is based. About 40 per cent live in consistently sunny tropical countries and at least another fifth inhabit areas with high levels of solar energy. Almost all of those without any electricity today live in areas with good potential for PV. In most of these regions, solar arrays combined with overnight battery storage can give householders reliable twenty-four-hour energy.

In northern countries with long winters, overnight storage isn't enough. We will need to take surplus energy for solar (or from wind) in times of abundance and turn it into renewable gas and liquid fuels. This will enable us to meet the need for power during the months when solar and other renewables are not producing enough.

The chemistry of producing zero carbon gases and liquids is not complicated and there are innovators already producing hydrogen from electrolysis and reacting this with CO_2 to create useful energy sources that can be stored in the existing oil and gas networks. In most countries this will give enough storage to provide for the needs of the entire winter.

So the ingredients for the switch to solar are all in place around the world. We have a cheap source of power, the software to manage the grid in the face of unpredictable electricity production, increasingly inexpensive daily storage in the form of batteries, and new ways of converting light into energy-rich gases and liquid fuels to provide the long-term reservoirs of power that a world reliant on solar PV needs. We have the ability to give everybody what Bill Gates wants: 'cheap, clean energy'. The only conceivable alternative – a new generation of nuclear power plants – seems far more expensive, riskier and much slower to build than a wholehearted switch to solar.

But will the transition happen? It needs vast amounts of capital to build the PV capacity to provide all the world's power, peaking at perhaps 3 per cent of annual global income. It requires commitment from governments and their electorates to live with the inevitable difficulties of switching away from fossil fuels to non-polluting alternatives and to back the research needed with the billions of dollars necessary to keep on pushing costs down. Lastly, and perhaps most importantly, it entails the biggest companies still tied to fossil fuels –whether it be the global oil companies, electricity utilities or car manufacturers – deciding that their future is unambiguously based around renewable energy sources. They will have to pivot towards the sun as the main source of the world's power and away from the carbon sources on which they created their businesses.

A year ago, I might have said that overcoming this last hurdle was impossible to imagine. The industrial dinosaurs of the twentieth century all seemed determined to resist THE SWITCH for as long as they could. It looked as though they would die defending their destructive business models, taking trillions of dollars of capital down with them. But as I revised the draft of this book in spring 2016, the evidence began to build that the monolithic determination to resist the inevitable was ebbing away. More or less explicitly, many of the main utilities began to voice a new strategy of reconstructing their activities around renewable energies.

Within the space of a couple of weeks in February, some of the world's largest energy businesses confirmed their move into the new world of low carbon energy. AGL (Australia's single largest emitter of CO_2), Engie (a $100 billion French utility operating in 70 countries) and the UK's electricity trade association all announced strategies of shifting away from coal, gas and oil into a future based on renewables, energy efficiency and smart management of the grid. Australia's AGL sold its entire gas business while Engie got rid of coal power stations. The UK's utilities said they wanted to bring an end to the use of coal for generation and asked the government for clear policy frameworks to enable them to move their capital into the new low carbon economy. Utility companies in some US states, such as Oregon, have switched sides and now lobby their legislatures sitting alongside the activist groups seeking to promote ever more demanding targets for renewable energy. These new corporate converts have come to recognise that the move to low carbon sources of energy will happen without them if they don't get involved early.

Will the large oil companies follow the electricity and gas utilities? They are even more embedded in the last century, and seem blind to the opportunity to lead the switch to renewable gas and liquid fuels. But economics will eventually swing even them towards solar-based energy. The large shareholder-owned global companies spend about $200 billion a year on exploration and production for hydrocarbons. On my calculations the energy produced – in the form of cubic metres of natural gas and barrels of oil – from this vast yearly investment is now less than the amount that would be generated by solar panels on which the same amount of money was spent. That is not a credible business plan. And indeed some of the oil chiefs are aware of it. In September 2015 Shell's CEO said that solar would become the '*dominant backbone*' of the energy system. If even Shell is saying this, the arguments for THE SWITCH are truly irrefutable.

In a recent book, the founder of PayPal, Peter Thiel, wrote that 'most peoples throughout history have been pessimists'. He might have added that pessimistic cultures such as ours find it difficult to deal with intractable problems, such as turning away from carbon-based energy. But now, I would argue, we can become optimists about our ability to address climate change at the same time as providing sufficient energy. And not just for the economies of the richer half of the world. PV provides a realistic prospect of delivering electricity to the billions of people who either have no access to power, or whose supply is fragile and intermittent. THE SWITCH will benefit all of us. It seems no exaggeration to say that in the developing world it will have a similar impact to mobile phones, as countries with little energy infrastructure leapfrog the model of national grids and carbon fuels.

A note on numbers and references

The APPENDIX contains some detail on the key units of measurement in the world of solar and other energy sources. If you are not familiar with quantities such as kilowatt hours or terawatts, it may be helpful to scan these pages first. There is nothing remotely difficult there.

The vast majority of REFERENCES for my research on this book are online and I have therefore provided links to these on my website, for ease of use. Please see:

www.carboncommentary.com

Chapter 1
Inevitable solar: the experience curve

No talk about energy is complete without the speaker noting that the sun gives us enough power in less than an hour to provide the total energy needs of the world for a year. This is not quite right. It actually takes about ninety minutes for enough solar energy to reach the surface of the earth. A small fraction of this energy is in the form of ultraviolet light (the sort that burns you when you are outside for too long) but most comes as infrared radiation (the kind that warms you when you stand in the sun) and visible light.

The numbers are simple. Across the whole year, an average of about 90,000 terawatts of solar energy is getting through the atmosphere and hitting the planet's surface. (The earth's annual orbit isn't precisely circular and slightly less energy arrives when our planet is a little further away from the sun in January. To compensate, July gets more sunlight.)

How does this 90,000 terawatts of solar energy compare to our needs for power? The world today is using an average of 15–17 terawatts at any one time. Just to be clear, that is the total running energy demand, not the amount consumed in any particular period. Divided over the human population of around 7.4 billion, this represents just over 2 kilowatts per person of continuous energy consumption, about the same as an electric kettle working all the time for each of us on the planet. Most of this power is generated by the burning of fossil fuels around the globe.

The world's current need for 15–17 terawatts is about one six thousandth of the energy that reaches us from the sun. 1/6000 of a year is about an hour and a half.

So, in the broadest terms, the earth's surface will always receive enough useful light and infrared energy every hour of the day to give everybody all the power that they could conceivably need. We just need to capture it cheaply and get it to everybody who needs it.

The plateauing of energy demand

Of course, the figure of 15–17 terawatts is not fixed. Global populations are set to rise and, currently, some 18 per cent (or 1.3 billion) of people are without any form of electricity. At some stage the world's poor must get access to power, and many others will buy cars for the first time and install energy-guzzling air conditioning. 2050's energy needs may be a multiple of what we use now. Can we still be confident that solar energy can give us enough power then?

The first point to note is that energy needs are now almost flat in the world's richest nations. In fact, total energy demand has been flat in the richer countries for over a decade. This surprises most people but, essentially, the wealthier nations have got as much power as they will ever need. Even if the industrial economies double or quadruple in size, energy use will probably not rise.

This seems counter-intuitive. The rise of the modern economy, whether in Europe, North America or Asia, was strongly linked to huge increases in fossil fuel consumption. Until quite recently, Chinese energy demand was growing as fast, or faster, than its annual economic output. There was a time not long ago when the country was said to be building two coal-fired power stations a week.

That era has passed for industrial countries, including China; our future growth will come not from using millions of tonnes of energy-intensive steel but from making smaller and smaller semiconductors. In the developed world we have largely got the stock of metals and concrete that we need and our energy demands are getting more modest, particularly as we get better at recycling everything.

In Britain, for example, the total amount of energy used peaked in the years just after 2001, has fallen by an average of 2 per cent a year since, and is now lower than it was in 1970. At the peak, the average person was responsible for about 5 kilowatts of running energy use (two kettles boiling away all the time, twenty-four hours a day). This number might seem absurdly high but when we are driving a car, for example, we might be using 40 kilowatts of power. and, even if we only drive it for an hour a day, that adds about 1.6 kilowatts to your average 24 hours running energy use. Today, the per person figure is about 4 kilo-watts and falling fairly consistently at a rate of several per cent a year. Some of this energy is consumed in the home, about 30 per cent in the form of motor and aviation fuels, and some by private and public institutions providing goods and services to us.

Is the fall in British energy consumption from 5 to 4 kilowatts over the past decade or so typical of other richer countries around the world? The UK's rate of decline is faster than in most places, but the EU as a whole managed a reduction of 8 per cent between 2005 and 2013. In the US, demand fell by about 6 per cent in the eight years to 2012. The energy use of the advanced economies appears to have peaked.

I'm going to guess that by 2035 the UK will need no more than 3 kilowatts per person. That's a decline of around 25 per cent on today's levels, requiring a reduction of just over 1 per cent a year, a far lower rate than the reduction seen in the last decade. Switching the population to electric cars and buses will provide a substantial fraction of this. Instead of using internal combustion engines, which are about 25 per cent efficient at turning energy into motion, we'll be using electric motors with efficiencies of up to 90 per cent. Similarly, many of us will be using electric heat pumps to heat our (newly well-insulated) houses and using far less energy than the gas boilers of today. All our lighting will be provided by highly efficient LED bulbs.

It seems a reasonable hypothesis that 3 kilowatts per person can provide a decent standard of living, wherever you are in the world. So I am going to assume that figure is roughly what the 9–10 billion people on the planet in 2035 will need to ensure access to light, heat, electricity and transport, and for businesses and government to supply goods and services.

Interestingly, BP agrees with this central idea that energy demand is close to a peak in rich countries. Its annual forecast for 2015 suggested a rise of a third in global energy demand by 2035 – but all from outside the US and EU. It predicts that power needs in today's rich OECD countries will barely change: China's energy growth will have slowed, and demand from the rest of the world will also be flattening. Overall global energy demand in 2035 will be rising about 1 per cent a year, less than half the current yearly increase.

The chart below makes this transition clear. By 2035, BP says, the world will be using energy at a rate of about 23 terawatts, up from about 15–17 terawatts today. With a world population of 9 billion, that would give the average person around 2.5 kilowatts – almost as much as I think is needed.

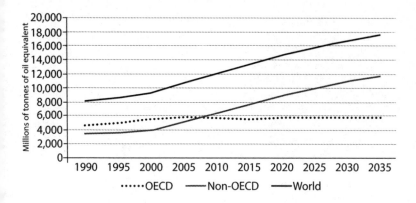

Energy demand in the rich OECD countries is flat and all future increases will come exclusively from other parts of the world.

BP's runing energy figure of 23 terawatts seems a reasonable estimate. But I'm going to be much more conservative and assume the planet actually needs 30 terawatts – almost twice today's level. I do this partly because I want to show that the switch to solar doesn't require us to stint on our use of power; a solar-based world will be a world of cheap and abundant energy. If we want to move globally to a low carbon energy system, without fossil fuels, then the big question is this: *can solar PV, storage and other renewable technologies provide 30 terawatts constantly, twenty-four hours a day?*

The availability of renewable energy

The solar resource available to us is enormous and is thus the obvious provider of our power. Whereas there's 90,000 terawatts of solar energy to use, the total amount of power in the wind is less than one per cent of this (around 870 terawatts). The next most important potential source is biomass – plants and trees – at around a thousandth of the amount of solar energy. And the

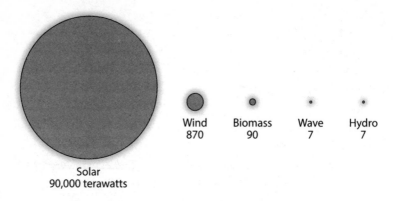

A clear winner: annual global energy capacity for solar (90,000 TW), wind (870 TW), biomass (90 TW), wave (7 TW) and hydro (7 TW).

potential for hydro-electric power or energy from waves across the world is another order of magnitude less.

The direct power of the sun is almost a hundred times as important as all other renewable sources put together. And indeed, all these other potential sources of renewable energy are indirect products of the sun's rays. Wind's chief function is to carry heated air from the hot tropics to cooler higher latitudes. Plant and tree biomass is produced as a result of photosynthesis that has used the sun's energy to create complex carbohydrates with a high energy content. Waves are the product of offshore winds. Hydro-electric power arises from the condensation of water into clouds which then rain on to high ground.

The other sources will, however, play important roles. Wind turbines will be really helpful in meeting the world's 30 terawatt need, although we cannot realistically expect to capture more than a small fraction of the 870 terawatts available. Wind blows over oceans, remote deserts and inaccessible mountains. Harvesting more than a few terawatts of power is going to be almost impossibly difficult.

Wind's role will nevertheless be a vital complement to solar in many locations, such as for the UK, in offshore locations around the edges of the North Sea, where the winds blow at greater speed and reliability than anybody realised even a few years ago. Biomass will be similarly useful to meet demand when the sun isn't shining, though we have deal first with the conflict with food growing (Chapter 4 looks at how this might happen). Hydro-electric can play a smallish part, too.

Nuclear power might help the move away from carbon fuels. However, problems of cost and construction difficulties are holding back the next generation of power plants. A year ago I heard a distinguished nuclear engineer describe the EPR, the new nuclear design planned for Hinkley Point in Somerset and other places in the UK, as 'unconstructable'. And while the world has substantial reserves of uranium, it would very rapidly use them up if it ramped up nuclear without very substantial investment in fuel recycling (something that has so far failed to happen anywhere in the world).

If the world is going to move away from fossil energy, and decides not to use nuclear power, solar is going to have to perform the critical role. Everything else can only really fill niches, not form the mainstay of world energy supply.

Does the world have enough space for solar?

We can look at this question in a variety of different ways. The first is this. The world is receiving 90,000 terawatts of energy and will need to use as much as 30 terawatts for generation in 2050. In a reasonably sunny country, a solar panel generates about 20 per cent of its theoretical maximum. (This low figure is mainly because it is dark twelve hours a day.) So let's say we need 150 terawatts of solar generating capacity for our 30 terawatts of continuous power requirements.

A tightly packed solar farm occupies about two hectares (five acres) of space to generate one megawatt. There are a million megawatts in a terawatt, so we will need 2 million hectares for each terawatt of peak power. 150 terawatts of generation therefore requires 300 million hectares.

Is this a lot? Very approximately, it's 1 per cent of the world's land area: far from negligible but not impossibly large. It is also a figure that doesn't take account of the 50 to 100 per cent improvements that we can reasonably expect between now and 2050 in the efficiency of solar panels. Nor the harvesting of solar energy falling on the walls and roofs of buildings. Together, these factors should cut the area of open space needed to well below half a per cent of the world's land area. We might be able to manage with about a quarter of one per cent of the planet's land surface.

When energy geeks are talking about this point to general audiences, they often show a world map with smallish dots, representing the area needed to provide the world's energy. They put some squares in the middle of the Sahara or the Arizona desert or on remote parts of Asia. Actually, this is a cheat. We won't place our solar farms in regions like this. Most photovoltaic installations will need to be close to centres of population and large scale electricity transmission lines. Generally solar will need to be placed in crowded parts of the globe and we shouldn't try to deny this.

For a country with a high population density – like the UK or Bangladesh – a much larger fraction of total space will be needed than the 1 per cent suggested above. By 2050, Bangladesh, the world's most densely populated large country, might have 200 million people, or about 2 per cent of the world's total. To give all these people 3 kilowatts of running energy will require about 6 million hectares, or 40 per cent of the country (before the efficiency improvements or the use of the roofs and walls). This is clearly a problematic number. Even though Bangladesh will probably put panels on every single building, on floats in rivers and lakes and even above some crops, it will still struggle to meet all its energy requirements from solar electricity. It will need to import power from India and other countries and may have to hold its energy requirements below the 3 kilowatts per head level.

In the UK, the position is somewhat different. We will need energy for at least 70 million people on a land area of 24 million hectares. At current solar panel productivity in this cloudy country – with the amount of sunshine running at less than half the world average – perhaps 16 per cent of the land area would need to be occupied by solar to provide all the energy the country will consume. Large amounts of PV-collecting capacity will need to be mounted on roofs and south-facing façades, which will work relatively well because the sun is generally quite low in the British sky. And solar will need to be supplemented by the country's unusually large wind resource, particularly in winter. It will be a challenge.

These are the most difficult countries. Dense populations and, in the UK's case, poor insolation, make the solar switch harder than the average. Look at the US, however, and the position is very different. Here, even at today's efficiency levels, the total space needed for panels would only be about 3 per cent of the land area, even if no contribution at all came from wind power. The southwest US has some of the best sunshine in the world – and large parts have good wind as well – so the space requirements will be even smaller.

If the US can push its needs for electricity and other energy sources down to 3 kilowatts a person – a very much lower figure than current requirements – and we experience the expected improvements in the ability of panels to turn sunlight into power, then probably about 1 per cent of the US land area will have to be used for PV or other solar technologies. A large fraction will be on land of negligible agricultural value, although some will inevitably have high levels of unusual biodiversity or other important environmental features.

But can we rely on solar getting cheaper?

In the late 1960s the Boston Consulting Group (BCG) investigated the cost of producing semiconductors for one of its clients. The consultants found that every time the aggregate number of electronic devices that had ever been made doubled, the cost of each individual item fell by about 25 per cent. If, for example, producers made a total of 1,000 units and the cost was $10 each, then by the time a total of 2,000 had been made the cost would be $7.50 per unit.

Bruce Henderson, the founder of BCG, called this the 'experience curve' and his company later showed that the cost declines arising from this effect were pervasive across different industries, countries and time. Although the rate of cost reductions tends to be fastest in products that are manufactured using automated processes in large factories, the experience curve phenomenon can also be seen in office-based or even agricultural activities. The speed of the decline varies from fractions of one per cent for each doubling of total production to levels even greater than that observed in semiconductors. A recent note from BCG reported that: 'Hard-disk drives showed a cost decline of about 50 per cent for each doubling of accumulated production from 1980 through 2002, bringing the average cost per gigabyte from $80,000 in 1984 to $6 in 2001.'

The most astonishing decline has probably been observed in the cost of gene sequencing. The full cost of extracting the genetic instructions from the human sequence fell from about $100 million in 2001 to around $2,000 in 2015. That is roughly equivalent to a 60 per cent slope on the experience curve.

We'd be wrong to imagine that BCG consultants were the first people to notice the impact of the accumulated number of units produced on the cost of production. That honour probably goes to T. P. Wright, an otherwise little known aircraft engineer who produced an academic paper in 1936 called 'Factors affecting the cost of airplanes'. This paper gave his observations, based on real factory-floor experience in the US, of how the costs of making a particular aircraft fell as the total number of planes produced rose.

He called this phenomenon 'the learning curve' and even today his term is more widely used than BCG's equivalent expression. He found that the rate of improvement was approximately 15 per cent for every doubling of accumulated production, a number not too dissimilar to the 20 per cent declines typically seen in solar panel manufacturing over the last sixty years.

Nobody doubts that learning effects exist. They are too universal not to be real. Take yourself as an example and that item of flat-pack furniture you recently purchased from IKEA. It probably took you several hours to assemble and, if you are like me, you made several mistakes in the process. But if you bought a second piece of the same piece of furniture, you'd probably see a major reduction in the time taken (and an enhancement in the quality of the finished item). That progress would continue over further units, although the improvements would tend to get smaller. That's an 'experience curve' in BCG's language.

In the Introduction, I mentioned Peter Eisenberger, the man who led Exxon's investigation in the 1980s into the prospects for solar energy. He agrees about the underlying predictability of solar development and the role of 'learning by doing'. In an interview with *Bloomberg News* published in November 2015 and headlined 'Exxon predicted today's cheap solar boom back in the 1980s' Eisenberger reflected that with enough data on how investment and effort lead to progress, 'you can use it pretty accurately to forecast what might happen in the future'. Eisenberger was saying that today's costs of solar technologies could be forecast with reasonable precision using assumptions about how fast costs improve with manufacturing experience.

Explanations of why the declines occur so consistently are varied. Some suggesr that the dominant causes are the larger factories and bigger companies (so-called 'economies of scale') that occur as industry grows. Others point to the effect of many small and minor improvements in manufacturing processes, technological advances or the fact that humans are good at getting rid of irksome inefficiencies.

Bruce Henderson of BCG taught us that cost declines across different industries were remarkably closely tied to the total volume of a particular product that had ever been made. In apparent contrast, Moore's Law, a maxim invented by Gordon Moore, one of the founders of chip-maker Intel, postulated in a 1965 paper that the maximum number of transistors on an integrated circuit would double every two years, while the cost of the chip would stay the same. Such an improvement would mean an approximately 41 per cent rate of reduction, year after year, in the cost of making transistors. In the way it is framed, Moore's Law appears to suggest that the passage of time drives the improvements. But the cost reductions that we have seen could equally be explained by a steep experience curve and the very rapid expansion in the numbers of transistors manufactured.

For the last half century commentators have questioned whether manufacturing cost improvements can continue for electronic components. So far, they have been wrong time after time. Looking out to 2020, experts now suggest binding physical limits to the number of transistors on a chip. At that point Moore's Law will cease to work.

But nobody yet sees comparable boundaries to the fall in the cost of collecting solar energy. Analysts attempting to calculate the percentage fall in cost from each doubling in accumulated global output of solar panels have almost all arrived at a figure of 20 per cent and the PV industry named this frequently observed, apparently highly predictable cost decline 'Swanson's Law', after the CEO of a solar PV company. The chart below shows the average cost of a watt of solar PV generating capacity on the left hand scale. The bottom axis is a measure of the total volume of PV panels produced, expressed

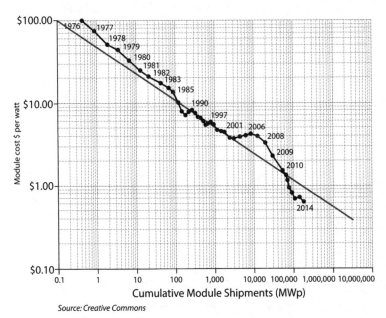

Source: Creative Commons

Swanson's Law tracks the cost of solar PV modules. The cost of solar has fallen predictably and consistently from 1976 to today.

in megawatts of peak output (MWp). Each point on the line in the chart measures the position at the end of each year since 1976. Although the speed of fall varies from period to period, the slope matches a 20 per cent experience curve over the course of the last forty years.

Twenty per cent is a slightly slower rate of decrease than BCG calculated for the early semiconductors half a century ago but still a strikingly fast rate of decline – and the figure has been standard wisdom for some time. Back in 2004, a group of Dutch and German researchers exhaustively discussed the evidence for the experience curve for photovoltaics. Their conclusion was clear-cut. The solar experience curve existed and the rate was somewhere between 20–23 per cent. If anything, they concluded, it was increasing over time. A couple of years later Travis Bradford, an American fund manager, calculated the decline at around 18 per cent for each doubling in the period 1980–2003. (Bradford's work is the earliest I have seen that asserts unequivocally that the resilience of this figure over more than three decades makes the eventual dominance of solar power a near-inevitability).

Other studies have compared the PV experience curve to the declines seen in the cost of onshore wind power. Here, the degree of cost reduction appears to run at about 8 per cent for each doubling of installed capacity. This means that although turbines in the windiest onshore locations offer electricity more cheaply than solar today, this will change. In the longer term solar seems certain to be less costly.

Other sources of power, such as nuclear, appear to have very slow declines in the expense of construction, at least in the last couple of decades. In fact, almost every study shows that the recent costs of nuclear power stations have increased sharply. This is probably because each new generation of power station is more complex and difficult to construct than the last.

More money is spent on safety. Tony Roulstone, a nuclear engineer from Cambridge University, says that the learning curve for nuclear doesn't have a downward slope also because each construction site uses different workforces and construction engineers who don't learn from each other's experiences. There is very little learning by doing.

If the world is going to use nuclear power instead of solar, it will have to find a way of getting plant construction on to a similar experience curve to solar. Since the first civilian nuclear power station in the world was built in Lancashire in the mid-1950s, this has never happened. Even France, which put in place the most sustained programme of nuclear power construction ever seen, saw sharp increases in costs over the two decades of construction. Prices approximately tripled over the period for each unit of electricity generation capacity.

Returning to solar, what does Swanson's Law mean for the future? As of late 2015 the world has manufactured something over one billion individual PV panels. Were the sun were shining on all of them at the same time, they would be continuously producing about 240 gigawatts of electricity, almost ten times the UK's requirement in the middle of a summer day. By the time the planet has two billion panels, the cost will be 80 per cent of today's level – if the learning curve forecast is correct. The initial price of the PV system is by far the most important determinant of the cost of the electricity it will produce. If a 10-kilowatt installation on a school roof costs £10,000 today, but £8,000 in two years' time, and nothing else changes, then the implicit cost of its electricity will also have fallen 20 per cent. The learning curve in manufacturing PV kit drives down the price of solar electricity.

The Swanson figure of 20 per cent reductions for every doubling of accumulated production of solar panels is a vital number. But it is only part of the equation. The other crucial element is the rate of expansion of manufacturing volumes. During 2015, the world manufactured enough panels to produce about 55 gigawatts of peak power, adding to the 175–185 gigawatts already installed. (Almost all the panels ever made are still generating electricity.) So in 2015 the world's total accumulated PV production rose by slightly more than 30 per cent. Swanson's law predicts that this reduced the cost of PV by about 8 per cent. (The maths here isn't entirely simple because each bit of extra accumulated production reduces the cost of PV by a gradually smaller amount.)

The rate of future fall in costs depends on the combination of Swanson's Law and on the speed of expansion of PV manufacturing. More precisely, cost reductions will depend on what percentage the new capacity adds to the total accumulated production each year. Currently, expectations are for installations of at least another 65–70 gigawatts globally in 2016, adding another 25 per cent to cumulative production, and reducing costs by a further 6 per cent or so.

This may well be far too conservative; installations of PV have grown on average by a fairly consistent 40 per cent a year for the last few decades and US, Chinese and Indian installations were sharply accelerating as we entered 2016. One recent prediction from industry analysts suggests the US alone will put 16 gigawatts on the ground and on roofs during 2016. And in the UK, where on the first weekend of April 2016 power from solar exceeded that from coal, it is amazing to note that *99 per cent* of that PV capacity had been deployed since May 2010.

What do cost declines mean for PV?

Photovoltaics currently supply less than 2 per cent of the world's electricity but such is the awesome power of compound growth that a straightforward continuation of the average 40 per cent rate of growth over the last half century would see about 150 terawatts of solar PV installations by about 2035. That would provide enough for all the world's energy needs – not just electricity.

But this is not the likely shape of the upward curve. More probable is that growth shades off slowly as PV grows. Most products and services that eventually become universal (such as access to the internet in developed countries or mobile phone ownership in the newer economies) follow a pattern often described as the 'S curve'. With the help of Professor Nick Jelley of Oxford University, I developed some arithmetic that allows us to model how PV might actually follow an S curve between now and about 2040. One possible pattern is shown below.

This is a description of what might happen and is certainly not a prediction. Nevertheless, I think it illustrates some important hypotheses. In this model, a 40 per cent growth rate continues for a few years and then begins to fall gradually, and later with increasing speed. By the late 2030s – as almost all

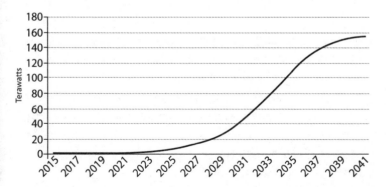

The S-curve. Forty per cent annual growth, slowing as saturation arrives, gives the world an energy system entirely based on solar within twenty-five years.

energy needs are fulfilled by PV – the amount of new capacity installed each year falls to almost zero. Slower growth rates in later years would mean, of course, that it will take longer to move to near complete reliance on solar energy. Nevertheless, the complete transition still occurs by 2041.

If, instead of 40 per cent growth, we see 30 per cent increases each year then it takes until mid-century to get to complete reliance on solar. An even lower rate of twenty per cent would mean it will be at least fifty years before that target is reached. Importantly, that 20 per cent growth rate would mean that the world would lose any hope of achieving a target of less than two degrees of global warming. But a continuation of 40 per cent growth may enable us to achieve close to zero emissions sufficiently soon to avoid the worst impact of climate change.

If we assume that an experience curve of 20 per cent slope continues to apply – and I can see reason why not – then the cost of PV in a world of 40 per cent yearly growth falls from about $1.20 a fully installed watt in late 2015 to as low as 15 US cents in 2041. (The faster the growth, the quicker the cost of solar declines, if the experience curve continues.) The crucial impact of this is that the implicit cost of electricity produced by solar PV in 2015, assuming an interest rate of 6 per cent and average amounts of solar radiation, is about 7.5 cents per kilowatt hour and will be little more than 1.5 cents by 2041. No other technology is likely to come remotely close to achieving this level. To labour the point, if PV is competitive now in the sunnier places around the world and falls to about one fifth of today's cost in 2041, it would be strange if it did not take over the world energy system.

Whether these individual numbers are accurate forecasts or not is not particularly important. The crucial thing is that a continuation of the trends of the last decades (40 per cent

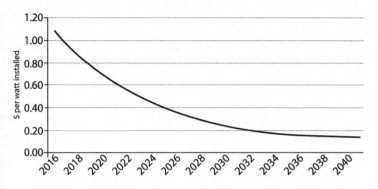

If current trends continue, the price of installing solar PV will fall to less than a sixth of current levels by 2041.

annual growth and a 20 per cent slope on the experience curve) can get the world free of fossil fuels within about thirty years. If world energy demand doesn't rise as fast as I have assumed then the process of complete decarbonisation would be even quicker.

The even more surprising conclusion is that the net cost of this to the global economy may be less than zero. As solar increases, the amount of money that the world needs to spend on fossil fuels falls. There is simply less need for coal, oil and gas as PV replaces them as the source of energy. I compared the total cost of adding enough solar power to completely replace fossil fuels over the next few decades with an alternative scenario that sees all renewables growth stopping tomorrow and the world needing more and more fossil fuels to meet its needs. If PV increases sufficiently rapidly, the world actually saves money by 2041. The creation of an entirely fossil-free global society is costless over a twenty-five year period.

At today's depressed prices, the world spends about $2.2 trillion on fossil fuels every year. If total energy demand rises from around 15–17 terawatts to about twice this level, then the total expenditure on fossil fuels between now and 2041 will be just over $81 trillion (which is about the total size of annual world GDP today). The total cost of installing enough PV by 2041 to meet the entire energy needs of the world will be slightly less.

It may even be better than this. In my calculations, I assumed that fossil fuel costs stayed at the very low levels of early 2016, with oil at about $32 a barrel and coal and gas also very cheap. At higher levels of fossil fuel prices, the benefits of switching to PV as fast as possible would be greater.

The graph below shows how expenditures might evolve year by year. The thick line estimates the total cost of installing PV and buying sufficient fossil fuels to meet the energy demand unfilled by solar energy. The dotted line shows how total fossil fuel costs would rise if there were no PV growth at all as the total global need for energy increases.

The PV line rises to a peak of around $5 trillion about 2032. It then represents a combination of around $3 trillion of solar PV installation costs and $2 trillion of remaining fossil

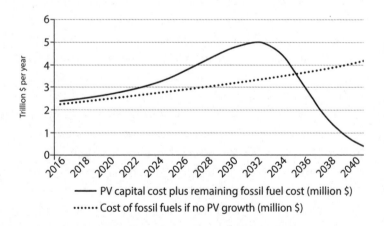

Moving to an energy system based on PV costs more over the next few years but then saves huge sums each year.

fuel expenditures. This compares with about $3.3 trillion of carbon fuel costs if PV installations stopped completely tomorrow. So in 2032 THE SWITCH to PV is costing the world about $1.7 trillion more than a fossil fuel future. This is the worst year. By 2036, the aggregate of new PV costs and the remaining carbon fuel expenditure is less than the cost of a no-PV world for the first time. By 2041, the fossil fuel dominated scenario costs $4 trillion a year compared to almost nothing for a PV-based economy.

It's worth pointing out that if fossil fuel prices return to their levels of 2013 then even the worst year of 2032 would see a lower cash bill from the switch to PV than from unrestrained growth in conventional fuels. In a world of higher oil prices than today, there might not be a single year in which the transition had a net financial cost.

I think the conclusion is clear. The world will benefit in cash terms, without even considering climate or air pollution benefits, from the move to renewable energy. It takes until 2041 to drive fossil fuel costs to zero but by that date the world has, over the full twenty-five-year period, actually saved money even at the low oil prices of $32 a barrel used in these calculations. By contrast, if PV grew more slowly than 40 per cent a year, not only would the world both have magnified its problems of climate change and air quality, but it would also not have saved money. For example, a growth rate of 30 per cent a year would result in the world being almost $4 trillion less well off than if PV continued on a 40 per cent growth curve. In short, the world's financial interests lie in pushing PV growth hard so that it continues to grow at its historic rates. Even if you believe climate change is an invention of the world's scientists and that air pollution doesn't result in premature deaths, PV makes straightforward and hard-headed financial sense.

It may be obvious but the key reason for all this is the experience curve. The more rapid the growth of output of an industry, whether it be semiconductors, mobile phones, LED lights or solar panels, the faster the rate of cost improvement. As a result, I suggest that if the world had to have just a single energy policy priority, the rational thing would be to back the continued rapid growth of PV panel production and installation to get the benefits of the experience curve as quickly as we can. (Later on in the book, I'll argue for a focus on further research into the improved conversion of solar light into liquid fuels as a second priority.)

Perhaps, like me, you find the extraordinary impact of fast growth combined with a steep experience curve difficult to comprehend. How can it be possible to get from 1 per cent of all energy requirements to close to 100 per cent in twenty-five years? Human psychology is very bad at understanding the long-term effects of a consistent percentage increase. This is perhaps best illustrated by the old story of an Indian king who offered his servant a prize for inventing the game of chess. The servant's claim was modest. He wanted some rice. All that he required was one grain on the first square of a chessboard, two on the second, four on the third, and so on. The king agreed, thinking the servant was a fool for asking for so little. His views started to change as the doubling from one square to the next started to require large quantities of grain. The king never reached the 64th and final square because by that time the amount of rice would have exceeded total world production by a factor of 1,000. Like most humans, he'd been unable to foresee the impact of continuous compound growth.

Some futurists note the importance of this deficiency in our psychology. They refer to it as the 'second half of the chessboard' problem. While we can envisage the impact of the doublings on the first few squares from a very small beginning, we do not have the ability to comprehend what will inevitably happen as the exponential growth continues. Solar power is still in the first few squares of the chessboard and none of us can yet see that continued rapid growth will eventually completely change the shape of the world energy system.

And we don't only have a blind spot about compound growth. Another weakness is that we cannot comprehend the impacts of two forces acting at once and strongly reinforcing each other. What has happened in solar PV over the last half century is that increases in accumulated production have caused major declines in the cost of manufacturing. These cost reductions have then resulted in lower prices and, inevitably, higher volumes of orders – 'a virtuous circle' as economists like to call it.

Even experts have problems with forecasting the speed of cost decline. In the 2004 Dutch/German study I mentioned on p. 22, the researchers told their readers that the experience curve was real and would probably persist in cutting PV prices indefinitely. But they went on to reduce expected industry growth rates to only 15–20 per cent rather than the 40 per cent annual rates experienced up to that point. As a result, they said that 'reaching break-even for the wholesale electricity market will most probably not happen before 2030'. For sunnier countries, that assertion is already incorrect.

It was the same for transistors, the building blocks of the electronic circuitry that now dominates our lives; lower prices meant that the range of financially viable applications increased year after year. Now we are actively talking about incorporating integrated circuits into almost everything that is manufactured because they are so cheap.

The volume of transistors that are shipped each year continues to rise, further driving experience curve benefits and cutting costs yet more. One analyst recently estimated that the number made is rising by about 65 per cent a year, meaning that over the next eighteen months more will be made than have been made in their history. The cost has gone down more than a thousand billion times since 1955, largely due to this expansion.

Dan Hutcheson, the head of industry research firm VLSI Research, made a comment about the growth of transistors that is relevant to PV:

> This steady, predictable decline in prices was a self-reinforcing gift. Because electronics manufacturers could depend on Moore's Law, they could plan further ahead and invest more in the development of new and better-performing products. In ways profound and surprising, this situation promoted economic growth. It has been the ever-rising tide that has not only lifted all boats but also enabled us to make fantastic and entirely new kinds of boats.

To compare the likely progress of PV over the next few years with transistors is not fanciful. Solar panels use some of the same principles of physics as transistors and improvements in manufacturing costs will come through similar processes of miniaturisation. The major solar PV manufacturers are all aware of the need to invest in performance improvement, knowing that if they do not spend they will be left behind by the cost improvements captured by their competitors.

I cannot see any reason why the experience curve in solar PV will stop. The world will use lighter and cheaper materials for the panels – in fact they probably won't be panels in twenty years but rather sheets – and the electronics to connect to the electricity grid (using huge numbers of ever cheaper integrated circuits, of course) will be getting smaller and more efficient. Chapter 3 has more details of how the technology of solar photovoltaics will improve.

Doyne Farmer, a mathematics professor at Oxford University, has made a study of the reliability of experience curve trends. If we know a particular technology has fallen in cost by 20 per cent for every doubling of accumulated production, how certain can we be that the trend will continue? Doyne and his colleague François Lafond did their work based on dozens of experience curves for different products and showed the underlying reliability of the continued decline in cost. If a product is on an experience curve of a particular rate, it tends to stay at that percentage. They showed, based on statistical analysis of the history of other commodities, that there is a small percentage chance that the PV cost decline will be halted or interrupted but that the odds are very low. Much more likely is that the decline in costs will continue at the same historic rate. They suggest that the most likely module cost in 2030 will be about 13 US cents per peak watt, about one sixth of the cost in 2013.

Not just falling costs for solar; the price of alternatives is tending to rise

Writers should probably avoid the phrase 'tipping point', when referring to a sudden speeding up of the transition between one state of affairs and another. The expression usually hides a lack of clarity about what is really going on. However, it is exactly what seems to be happening with solar power. Having struggled for decades, with support from eccentrics and specialist users, solar has now become the obvious, almost unthinking, choice for those adding new power capacity in many parts of the world. In some of the months of 2015, solar and wind together provided almost 100 per cent of the additions to the US grid, for example.

When another solar farm is added to the grid, or a few new panels put on someone's domestic roof, electricity utilities know one thing with absolute clarity. In almost all circumstances every single kilowatt hour of power produced by the new PV system will be used. Electricity from PV is free to produce and so even if the market price of power falls close to zero, solar power will still be shipped out into the grid. Once installed, PV (and wind-generated) electricity will always be able to undercut all other sources of power.

However low the price of gas falls, it will never be cheaper to fire up a gas-fired power station than to take electricity from a solar farm. As solar penetration rises, operators of fossil fuel plants are therefore only too aware that the total amount of electricity they will be called on to supply will be smaller and smaller each year as solar grows.

Put yourself in the position of an electricity utility trying to decide whether or not to spend upwards of a billion pounds to build a new Combined Cycle Gas Turbine (CCGT) plant. It varies a bit around the world but generally power stations compete in a market. Buyers of electric power – usually companies that then sell electricity to homes and businesses – will purchase power from whatever source offers the lowest price for the next hour. If solar or wind are producing enough electricity, coal and gas power stations will be idle, not even earning the money needed to pay staff or maintenance costs, let alone the interest on the money borrowed to construct the plant. So the company facing the decision about whether to build the new CCGT plant needs to forecast whether it will earn enough to cover its costs and make a profit.

In front of the investment decision-takers will be a report containing the two key numbers: a projection of the number of hours a year the power station will work over the twenty-five or thirty years of the plant's life and an estimate of the price it will get for the electricity it generates when it is working. The first of these numbers might be 2,000 hours, or it might be 6,000, out of the nearly 9,000 hours in a year. A large number will mean that the plant is a valuable addition to the portfolio of generators providing electricity. It will be working most of the time. A limited number of hours will suggest that the power station will be used only when other sources of power are not quite providing enough.

On the same page will be an estimate of the price it will obtain for electricity during the hours it is working. The people round the table will start to look worried. If solar electricity is getting cheaper and cheaper – perhaps dropping in price by 8 or 10 per cent per year – and continuing to grow rapidly, then as time goes on both the number of hours that the CCGT will work, and the price it is able to obtain, will tend to fall. No one knows quite how rapidly the change will occur, but occur it will.

We can see the problem facing these decision-takers. The success of their gamble investing a large amount of their shareholders' money for the next twenty-five years or more will depend on whether the plant is going to be reliably used for profitable generation most of the hours in a year. They don't want their billion pound plant lying idle, losing cash. They cannot predict how fast solar will grow but they will have noted that even the CEOs of fossil fuel companies are beginning to say PV will eventually dominate. Trend-setters like California are already mandating 50 per cent renewables by 2030, leaving little space for gas or coal-fired power. A report from Bloomberg in October 2015 noted that the average utilisation of gas-fired power stations in the US had fallen from 70 per cent of the available hours in the year before to 62 per cent in 2015. Coal-fired power stations in China were used 51 per cent of the time in 2015, down from 56 per cent the year before. And this is only the start of the trend.

If these people sitting around a boardroom table were in the UK, they'd probably have access to the chart below, and it doesn't make enjoyable study. For a few brief months in late 2014 and early 2015 the UK was one of the world leaders in installing photovoltaics. Between summer 2014 and summer 2015, the amount of solar power almost doubled from about 4.4 to about 8.3 gigawatts (the numbers are vague because the UK government doesn't keep proper statistics).

The chart below shows the impact of this sharp growth on the need for power from gas and coal power stations at 1pm in the March–July period of 2014 and 2015. (I've used 1pm on British Summer Time because this is usually the time of peak solar PV output.) The UK had a total of about 44 gigawatts of coal and gas-fired power stations. In 2014, over half of this capacity was typically producing power at 1pm in summer. This average fell by 3.5 gigawatts in one year, reducing the

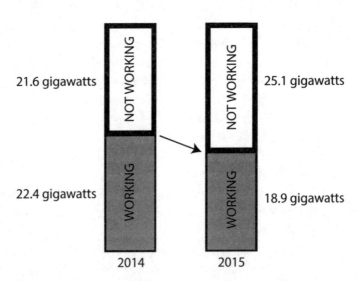

Patterns of working for fossil fuel power stations in the UK, summer 2014 and 2015

utilisation of available capacity from 51 per cent down to 43 per cent. As solar goes up, the amount of electricity needed from fossil fuel producers inevitably falls. If you own gas power stations with hundreds of people on the staff, this really hurts.

Over the summer of 2015, the 1pm need for fossil fuel generation fluctuated from below 10 gigawatts to a maximum of around 28 gigawatts. Lunchtime in the summer has quite a high daily power demand compared to the early and late parts of the working day but large numbers of gas and coal-fired plants never worked during the midday hours. There was a regular siesta for tens of fossil fuel stations across Britain. On one day in early June – an unusually sunny and windy Saturday – the need for coal and gas plants fell to just 9 gigawatts, implying that only one fifth of these units were working. It wasn't until the early evening that this figure rose to more than 10 gigawatts. This was just one day, of course, and the wind doesn't often blow strongly in June, but it shows what the future will be like for fossil fuel generating plants.

Bankers and investors note events like this and they are getting increasingly twitchy about putting their money behind new fossil fuel power stations. Their concerns are exacerbated by the long-term trend of declining electricity use, unrelated to the growth of PV. Even with these adverse background circumstances, however, sometimes a utility will commit to a new CCGT plant. Its reasoning will be that when renewables are unavailable at night or in periods of unproductive weather, the market price will shoot up. So although the proposed power station will not work many hours each year, it will make more money when it does. Spikes in the electricity price can be short but sometimes reach astonishing heights. On one day in November 2015, the UK's National Grid was forced to pay a small number of generators about £2,500 per megawatt hour of electricity – over fifty times the going rate – because of an unexpected shortage of electricity. (This deficit actually arose because several fossil fuel plants all failed at the same time.)

However, someone in the room when the decision is made whether to built a new gas-fired power station will probably point out that these spikes may become less frequent in the future. 'In ten years' time,' the finance person may say, 'many businesses and households will be linked to a system that automatically reduces their electricity draw when the cost rises. This will tend to dampen the sharp upward moves. We cannot rely on high prices, even when solar (or wind) isn't working.'

He or she may get outvoted, but their concerns are reasonable. Today's variations in price are in part a function of the historic inflexibility of electricity demand. Most users, including homes and many small businesses, don't pay any more even when their utility has had to pay thousands of pounds for a megawatt hour during periods of electricity shortage. So there's little or no incentive on them to ratchet down their use. In the face of sharp increases in the wholesale price, utilities face inflexible demand. This will change dramatically over the next few years as companies like REstore (see p. 149) put sensors and algorithms in place that reduce electricity use in factories and warehouses when supply is tight.

When I spoke to REstore entrepreneur Pieter-Jan Mermans, he told me that before he set the business up he'd gone to visit a competitor that had been around for some years. That company's business, like his, revolved around cutting the electricity demand of major users at periods of grid stress in return for payments from the operator of the grid. They did this work using old-fashioned fax machines and telephone calls. And they could never be completely sure that the electricity use at their customers had in fact been cut below what it would have been because they didn't have sensors or intelligent meters at the factories. That's less than ten years ago.

Now REstore has completely automated 'demand response' and hundreds of sensors at its customers measure the need for electricity every second. Demand can be throttled back quickly and easily. As a consequence the electricity industry is rapidly becoming like any other market in which tight supply causes buyers to back away, perhaps delaying their purchases or perhaps avoiding the need to buy altogether. The idea that electricity demand is inflexible and unchangeable is wrong, and electricity producers can no longer rely on being able to sell electricity into sharp price spikes at busy times.

As solar grows, along with wind and the other technologies discussed in this book, it will become more and more difficult to justify investing in fossil fuel power stations. Investors will get increasingly wary of backing new projects using gas or coal. Their insecurity will manifest itself in demands for higher financial returns, delivered quickly and before the solar revolution completely consumes the energy market. Those requirements will be harder and harder to fulfil. Solar will become cheaper – not without the hiccups of temporary price rises but generally much cheaper year on year – and the relative attractiveness of gas and coal will fall further.

Activist movements in universities and throughout social media will encourage the trend towards renewables and eventually it may become impossible anywhere in the world to build a gas-fired power station, let alone coal, in the face of public disapproval. In what may be a symbolic moment in early October 2015, a British developer admitted it could not find the finance for a long-planned CCGT plant in Manchester. Similar cancellations have been seen across the world, partly as a result of the rise of solar and other renewables and partly as a result of the worldwide flattening of electricity demand. Even in the US, the home of cheap natural gas as a result of the fracking boom, the installation of new CCGT plants is running at a fraction of the level of recent years. Coal is already in deep crisis with a string of bankruptcies in 2015–16, among them Peabody Energy, the world's largest coal company.

It may be too soon to say this, but the developed world may be close to collectively concluding that new coal and gas is financially unjustifiable in the face of lowered need for electricity and better availability and cost from solar PV. As PV grows, the financial rewards from investing in fossil fuel power generation continuously weaken. As that happens, the 'tipping point' will be a reality.

Governments have been slow to recognise this, not least because it produces some very uncomfortable conclusions about their capacity to determine the shape of the future electricity industry. In the UK, for example, the government minister in charge of energy policy, Amber Rudd, said in late 2015 that the national interest is 'best served by open competitive markets'. But no market can work effectively when a new entrant – in this case solar PV – can offer extra electricity at zero cost. If building new gas-fired power stations is necessary to provide a last resort source of electricity when other resources are unavailable, then they will have to be financed, or guaranteed, by governments. Or, as I suggest in the last chapter of this book, we'll need to develop sources such as renewable methane as the fuel for the gas-fired power stations some countries will continue to need when the sun isn't shining and the wind not blowing.

Chapter 2
Predictions – and the global switch

It's time to put my own hand up. I was at an energy conference in summer 2013 when one of the UK's most distinguished scientists told us that solar PV 'was so expensive that it would never form a significant fraction of the electricity supply'. He therefore proposed to ignore it in his talk. We all nodded agreement.

We wouldn't do so now. Rarely have opinions changed so rapidly or so conclusively about something this important. In an interview during September 2015 Ben van Beurden, the CEO of Shell, said that solar will become the '*dominant backbone*' of the energy system across the world. It wasn't completely clear from his comments when he expected THE SWITCH to occur, but he was in no doubt that the transition will happen. This was a startling new admission from the head of one of the world's largest fossil fuel producers.

A few weeks later one of the world's top climate scientists made a similar prediction. Professor John Schellnhuber of the Potsdam Institute, and an adviser to Pope Francis and Angela Merkel, said 'ultimately nothing can compete with renewables'. In December 2015, Nick Butler, former worldwide head of strategy for BP, was equally forthright. In one of his insightful columns in the *Financial Times* he wrote that 'advances along with falling costs promise to make solar the power source of choice in the 21st century'. Just eighteen months previously, also in the *FT*, he had declared: 'As things stand renewables will never break through economically … electricity produced from offshore wind and solar costs somewhere between 50 and 100 per cent more per megawatt hour than power from natural gas and, with some variations, will continue to do so for the next decade.'

Our slowness in accepting that solar PV is the logical choice for the world energy system partly stems from an inability to comprehend the predictability and consistency of the cost improvements. Danny Kennedy, managing director of the California Clean Energy Fund, and someone who might be expected to be an optimist, made the following remarks in an end of year summary in December 2015:

> It never ceases to amaze me how PV costs keep coming down. According to Bloomberg they were down again across the globe 15 per cent year on year. It is unparalleled in the history of energy uses to have a source keep getting cheaper and cheaper and cheaper year on year on year not by single-digit but by double-digit gains. I know I should know it by now, as it has been on this curve persistently for decades, but it still blows me away.

The International Energy Agency (IEA) was sceptical about solar for decades. For many years, it had ignored the growth of PV, thinking it irrelevant to the future of energy. It began to change its mind in 2014 and started actively following technical and financial developments, noting that solar had grown by an average of 47 per cent a year over the period 1990 to

2013. This somewhat half-hearted convert has now come round to saying that PV could be the biggest single source of electricity by 2050.

Even the people whose livelihood still depends on the fossil fuel companies have what I call the 'second glass' problem. Talking to them after debates or conferences they stick to the company line until the end of the first drink. They praise the potential of solar photovoltaics to eventually revolutionise global energy supply but slip in a few phrases that seek to undermine it or to suggest its growth will slow. Two years ago, it was the 'enormous' cost of PV that they'd mention. Now the worried sentences are about the impossibility of storing electricity and the problems of seasonal deficits in high latitude countries. But as they reach gratefully for the second glass of wine, their guard begins to drop. Their voice descends to a whisper as they surreptitiously take a quick look to see if anybody can overhear. 'It's obvious. Our industry [oil, gas or coal or even nuclear] is dying fast. We can't raise the capital to build new plants. By the next decade, PV will be a cheaper source of energy than any fossil fuel.'

Ben van Beurden of Shell admitted in the interview I quoted that solar will be the eventual victor, but being an oil-man he has to believe that fossil fuels have a few successful decades left. He went on to talk about the scale of the challenge facing PV between now and 2050:

> We will see the demand for energy double, and I do not think solar can grow from the one per cent it is now, to being the dominant force and thereby obviating the need for fossils fuel immediately, but it is going to be a multi-decadal transition.

He is right about the scale of the challenge. Today, it seems almost inconceivable that solar could grow to being the most important source of power within a couple of decades or so. However even fossil fuel experts recognise that not only is solar growing towards dominance but the storage revolution is arriving just in time. Nick Butler's December 2015 column in the *FT* went on to notice the growing belief in the world's ability to store electricity in bulk. Batteries aren't yet financially viable, he said, but that day is fast approaching. This was the first time I had ever seen someone from a fossil fuel background noticing the rapid decline in their costs. He referred to two reports from a bank and a credit rating agency that demonstrated strong belief in the rapid and widespread commercialisation of large scale energy storage. Butler, who has spent most of his working life working for an oil company, concluded: 'When serious and objective financial institutions start saying such things, investors and companies involved in the old energy economy would be foolish not to take notice.'

As with photovoltaics over the last few years, the speed in decline in electricity storage costs is surprising even industry observers. In Chapter 6 I explore the evidence that batteries are becoming cheaper in a very similar way to PV. In many countries they already match the costs of conventional ways of supplying short-term bursts of power into an over-stretched electricity grid.

The race to keep up with the speed of decline

Almost every single forecast over the last sixty years for the progress of solar power has been too pessimistic. Although the slope of the experience curve for PV has been well understood for at least a decade, no one seems to have predicted the continued rapid growth in installations around the world. As a result, cost expectations are systematically too high, even now. Before looking at the evidence of strong continuing momentum in solar installations around the world, it seems instructive to review some past forecasts. Protagonists for PV are often accused of naïve optimism; my purpose here is to suggest that the reverse is true. If the evidence from the last half century is any guide, THE SWITCH might happen even faster than anybody projects today.

In many cases recent projections for 2030 have already been achieved. Sometimes the figures for 2050 look within easy reach. The International Energy Agency's forecasts are a case in point. Their research institute's 'Technology Roadmap' for PV in 2010 suggested that the costs of installing a kilo-watt of PV in 2020 would be about $1,800. This would fall to about $1,200 in 2030. Their figure for 2020 was widely achieved around the world in 2013, and $1,200 was reached in many places by mid-2015, fifteen years early.

The resulting underlying cost of electricity was also strik-ingly badly estimated. The IEA gave us a figure of 21 US cents per kilowatt hour in 2020 in gloomy places like Britain or Denmark and 10.5 cents in the sunniest countries. In the 2015 UK competitive auction, developers were prepared to offer to supply electricity for the equivalent of about 12 US cents, less than 60 per cent of the IEA's estimate for five years later. The lowest bids from owners of PV farms in better irradiance areas around the world are now coming in at just above 6 cents per kilowatt hour. By 2030, the IEA said, the best countries would be achieving 7 cents per kilowatt hour. So, once again, the world is already achieving prices that are below projections for fifteen years' time.

It is not just the price of power, but also how much is being installed. The IEA estimated in 2010 that the world would have 200 gigawatts of solar power by 2020. This hurdle was achieved in early 2015. The annual rate of increase was expected to be about 34 gigawatts in 2020, a number beaten by late 2013.

In 2014, the IEA published a new roadmap for solar, concluding if current trends persisted that in 2050 'the best case will lead to generating costs lower than 5 US cents a kilowatt hour'. We haven't yet seen a number as low as that – the single cheapest completely unsubsidised contract agreed so far was set at just less than 6 US cents – but it looks likely to be a reality by 2020 at the latest. Thirty years ahead of time.

Lawrence Berkeley Laboratories (LBL), one of America's pre-eminent research institutions, collated the 2008 forecasts of major banks for the cost of PV modules in 2010, just two years later. On average, the banks' estimates were about forty per cent too high. Similarly, the predictions made in 2011 for the figures for one year later were out by 15 per cent. LBL's work shows how consistently wrong all forecasters, including banks and even industry bodies, had been. Predictions have almost always substantially overestimated the real numbers even just a few months into the future.

At the end of 2013, the respected Fraunhofer Institute in Freiburg projected that new PV installations would be generating electricity at less than 12 Euro cents a kilowatt hour everywhere across Germany by the end of the decade as solar panel costs continued to fall. (There's a short discussion of how we calculate the cost of solar electricity, based on the financial capital it requires to install the array, in the APPENDIX.) At its cheapest, the figure would be little more than 6 euro cents per kilowatt hour for big solar farms in the sunniest parts of southern Germany. This would be cheaper than generating electricity from burning lignite, the soft brown coal that is the mainstay of German power generation because it is so readily available. (It is also hugely polluting.) Perhaps as importantly, even Fraunhofer's higher figure of nearly 12 cents is less than half the price charged by German electricity companies to residential customers. By 2020, it would make financial sense for every householder to install PV on roofs, even in cloudy cities on the northern Baltic coast of the country.

Fraunhofer saw further declines beyond the end of this decade. It concluded that PV in southern Germany would produce electricity at costs well below all types of fossil fuel plants by 2030. In a good location, a new PV plant could be expected to generate power for around 5 euro cents per kilowatt hour, about half the cost of a gas-fired power station. Of course, this comparison is influenced by assumptions about the future price of fossil fuels but even a slump in coal and gas prices to nearly zero would still leave PV in a competitive position. And if the world does take aggressive action on climate change – not perhaps likely – the levies and taxes imposed on fossil fuels would further improve the competitiveness of solar.

Less than two years later, in February 2015, Fraunhofer wrote a report suggesting that by 2025 'the cost of producing power in central and southern Europe will have declined to between 4 and 6 cents per kilowatt hour'. It had moved its lowest estimate of the cost of producing electricity down and brought the dates forward. And it was at pains to stress the conservative assumptions it had used, stating 'technological breakthroughs could make electricity even cheaper, but these potential developments were not taken into consideration'.

A 2011 publication from the senior advisers to the UK government, the Committee on Climate Change (CCC), made its own estimates of solar costs in 2020 and 2030. Its Low estimate for 2020 was 14 pence per kilowatt hour and the High estimate almost double this. PV costs in the UK will always be higher than elsewhere because the same number of solar panels will produce far more electricity in a sunny country. Nonetheless, the Low figure was probably achieved by late 2013, while the CCC's Low estimate for 2030 (8.9 pence per kilowatt hour) was beaten during 2015.

Given that their projection for almost twenty years into the future was beaten within four years, the Committee revised its forecasts in late 2015. Rather surprisingly, however, it concluded that 2020 PV costs in the UK could be as high as 9.6 pence per kilowatt hour, actually above current levels, and by 2030 the cost might be 7.2 pence per kilowatt hour (barely 15 per cent below current levels). Although cost declines in PV have been remarkably consistent and large, the CCC simply could not bring itself to write down estimates that continued the trends of the last fifty or so years.

Compare nuclear power with solar PV. Between the two forecasts of 2011 and 2015, the CCC almost doubled its estimate of nuclear costs by 2025. It's not just that solar is getting cheaper but also that most other types of power production are getting more expensive. So even these highly cautious forecasters see the relative advantage of PV widening.

Confounding the optimists

The systematic pessimism about the future costs of PV has not just been confined to out-of-touch research institutions. Almost everybody has been guilty of excessive caution about the fall in the price of PV panels and, to a lesser extent, declines in the cost of all the associated electronics.

Even enthusiasts underestimated the rate of technical progress. Fund manager and university teacher Travis Bradford published a book, *Solar Revolution*, in 2006 that argued that PV would eventually allow the world to run largely on solar energy. Rather than directly estimate the future cost of electricity, he worked out what the construction costs of solar farms would be. If it costs less to put a solar array in place, the price of electricity from that installation will also fall. Bradford suggested that PV costs would decline at 5–6 per cent a year,

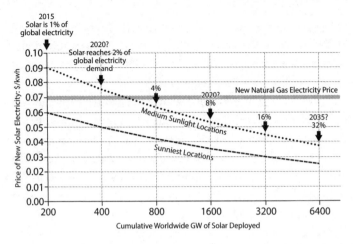

How cheap can solar get? These predictions show costs in sunny and medium sun locations as cumulative PV is deployed worldwide.

meaning that by 2015 the full costs of a solar farm would be down to $3 per watt, compared to around $6 in 2005.

Likewise, the futurist Ramez Naam made a prediction in *Scientific American* in 2011, noting that PV was riding down a clear cost reduction curve. Rather than estimating the price of a full solar system, he looked just at the price of the panels themselves, not the remaining kit that turns the output of direct current into AC power that can be sent into the electricity grid. He said the price of panels was about $3 a watt in 2011 and would fall by about 7 per cent a year. Therefore 2015 should see prices of panels of around $2.20. By 2030 or so, continued declines at this rate meant panels would fall to around 50 cents a watt.

These two writers are experts who had done their maths carefully. Both understood that solar PV farms were benefiting from the almost magical effects of 'learning by doing'. Moreover, Travis Bradford and Ramez Naam are partisans; they believe in the future of PV. Neither was remotely optimistic enough. The full cost of a solar farm, including all the cabling and electronics is now around $1.30–1.40 a watt in countries with large PV installation industries. That's less than half of Travis Bradford's 2006 estimate for 2015.

Panel costs have probably been falling faster than the rest of PV systems and the Naam estimate was even more wrong. By mid-2015, the wholesale price of panels, bought directly in large quantities, was about 50 cents a watt, and possibly a little lower. An estimate carefully made by a knowledgeable individual was wildly inaccurate. In fact, 2015 saw prices that Naam said wouldn't be achieved until 2030.

There have been just a few people who overestimated the speed of the decline. The visionary British solar entrepreneur Jeremy Leggett said in 2010 that UK solar would have declined sufficiently in price by 2013 to match the cost of domestic electricity bought from conventional suppliers. This was too optimistic; it was early 2016 before the UK government was reduced the subsidy for small scale PV down to almost zero, saying that it no longer needed significant support.

The purpose of the last few paragraphs is not to mock the respected institutions and people mentioned. Rather I want to suggest that we have all suffered from an extraordinary myopia. We have some sort of strange psychological flaw that makes it impossible for us to believe anything this good could continue much longer than it has done already. Study after study has showed rapid falls in the previous few years but the projections for the future are then manipulated to show much slower rates of decrease. It is almost as if the authors were simply afraid of being thought to be seen as cheerleaders gullibly supporting the solar industry's efforts to talk up its prospects.

But it now seems likely that the relentless progress in PV is beginning to erode the scepticism of the most hard-boiled analysts. By early 2015 the big international banks were beginning to circulate reports that stated solar was the inevitable future of power generation. Deutsche Bank's New York office wrote in February that solar had reached 'grid parity' in thirty of the sixty countries it had analysed. The bank also said that it expected a further 30–40 per cent reduction over the next five years, meaning that almost everywhere in the world will see new PV installations generating power for less than a recently built fossil fuel plant by 2020.

Citibank added its voice in August 2015. With more emotion in its language than is usual from investment banks, it wrote that low carbon electricity generation, particularly from solar PV, is overwhelmingly the best choice for a world threatened by climate change. Its charts showed sunny locations with costs below 5 US cents per kilowatt hour in 2020 and about 3 US cents by 2030. The downward curve of solar costs, crossing the flat lines of coal and gas generation in the next few years, is similar to all the other graphs from banks and research houses. It's almost become a visual cliché.

The evidence from developers on the ground

As 2016 dawned, the evidence for price reductions in solar PV swarmed in from all parts of the globe. Almost every day saw new and surprisingly low figures from different continents. These numbers, as expected, varied dramatically depending on the region and the amount of sun it received. But they were all hard figures, most often coming from tenders run by governments or power utilities looking for binding offers of electricity generation. Thus we can be reasonably confident that the prices offered by developers are a reasonably accurate reflection of the costs they face.

As well as erroneous forecasts for the costs of installing solar arrays, many have overestimated the likely price that would be demanded by PV farms for the electricity they generate. In the UK, the government's energy department made a forecast in mid-2013 that large scale solar PV would need to be paid about 12 pence per kilowatt hour to make it worthwhile to build a new farm in 2019. However, when it held an auction in February 2015 asking PV developers to offer prices for projects to be constructed a year later, it received offers as low as just over 8 pence per kilowatt hour. This was lower than the minimum cost the government's energy specialists had thought possible for 2030. It's also noteworthy that these bids were already significantly lower than the price agreed by the UK government for the electricity from a new nuclear power station at Hinkley on the western coast of England.

Governments are increasingly using open auctions similar to the UK's as the means by which they attract developers into building solar farms. Each participant offers an electricity price, expressed in cents per kilowatt hour, for power from individual locations. The past year (2015–16) has seen a sharp decline in the prices bid into these auctions everywhere around the world. When details are published of the prices that suppliers are prepared to bid to electricity companies to win the bid, the gasps of near-disbelief from the industry are almost audible to the outside world. One such shock was an auction in Dubai in late 2014 that produced an offer (without any government subsidy) of about 6 cents per kilowatt hour. Major Saudi companies were willing to install and run a huge 100 megawatt solar farm for less than it costs to produce electricity from gas. Dubai has extremely high levels of solar radiation but this bid was little more than half what observers were expecting. And it wasn't an outlier; another auction a few months later produced a bid that was slightly lower.

Published details of several bids in summer 2015 in the southwest US were as low as 3.9 cents, although this figure contains an element of subsidy from tax relief. The city of Austin, Texas bought 200 megawatts of solar power for less than 4 cents a kilowatt hour. A few days later Nevada's NV Energy agreed to buy electricity from a 100-megawatt farm for 3.87 cents per kilowatt hour, about a third of the rate that the utility had agreed to pay for solar power the previous year. A Bloomberg analyst noted: 'That's probably the cheapest PPA [power purchase agreement] I've ever seen in the U.S.' He wasn't talking just about solar offers, he was saying that this was the lowest offer to provide electricity from any source, including fossil fuels and well-established renewables such as hydro-electric dams. (Nevertheless, it may be that some wind PPAs have been cheaper than this figure.)

Tenders in sunny parts of the world may not always produce bids as low as some of these figures. Import tariffs on panels, difficult local conditions, high labour costs and requirements to upgrade the electricity grid will mean higher prices than in the sun-soaked southwest US or Dubai. But the general picture around the world is of rapid falls in the price that operators are willing to accept to install and run large solar PV farms.

The Indian switch

The shift in attitude towards solar has occurred almost simultaneously across the world. It is not just a feature of advanced industrial economies. In poorer regions, PV is seen as a chance in a lifetime to expand electricity availability.

Prime Minister Modi has set a target for India of 100 gigawatts of solar capacity by 2022 – a ten-fold increase on the country's 10 gigawatts today. Indian installations in 2016 will be more than double those of 2015. China has said it wants 150 gigawatts by 2018, a target that is widely seen as likely to be massively exceeded. In summer 2015, US Presidential candidate Hillary Clinton said her target would be 140 gigawatts by the end of her first term in 2020. That would provide about 8 per cent of US electricity.

India is a good example of worldwide trends as it begins its drive to get electricity to all of its population. In 2014 developers offered to build solar farms for an average of about 7 rupees per kilowatt hour. (That equates to around 10 US cents, or 7 pence, at current exchange rates.) Three state auctions in the third quarter of 2015 in Madhya Pradesh, Telangana and Punjab saw offers of just over 5 rupees. The Indian press openly speculated that these offers were too low to be profitable for their developers. But in November 2015 another round of tenders in Andhra Pradesh in southeast India resulted in a bid of 4.63 rupees (about 4.6 pence per kilowatt hour). This was for sites totalling 500 megawatts and was won by the US company SunEdison, the world's largest renewable energy developer.

Just before the Andhra Pradesh auction was completed, the accountants India KPMG released a detailed report on the state of PV in their country. 'We see solar power becoming a mainstay of our energy landscape in the next decade,' they wrote. Like everyone, they were still cautious about future solar PV bids: their best guess for auction prices was 4.20 rupees per kilowatt hour (about 4.2 pence) by 2020, only about 10 per cent less than the SunEdison November 2015 bid. Their figure for 2025 is 3.59 rupees, or about 3.6 pence.

What matters most in India is how well these numbers compare to electricity from inexpensive locally mined coal. KPMG says that the current cost of power from this source is about 4.46 rupees per kilowatt hour, only about 4 per cent below the November 2015 record low bid in Andhra Pradesh. But in a power station using some imported coal,

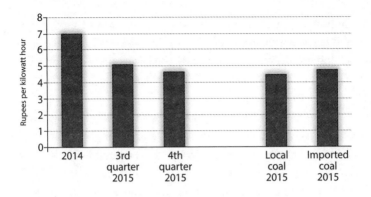

Bids for India's energy contracts, 2014–15. Solar has almost reached parity with both local and imported coal.

the accountants calculate, the cost would be higher than solar. In India, PV is therefore now directly competitive with some coal power stations and by 2020 it will be 10 per cent cheaper, KPMG conclude. They predict that the raw cost of solar electricity from big solar farms will be 3.5 to 3.7 rupees by 2025. (Around 3.6 pence.) If history is any guide, they are pessimistic.

These numbers are not complete. We also need to include the cost of getting the electricity to the final consumer. In many countries this penalises solar but not so in India, says KPMG. Many of the coal plants are hundreds of kilometres away from the centres of electricity demand so the relative attractiveness of solar is improved when electricity distribution costs are included. In fact, when the accountants have fully loaded the costs PV ends up as very slightly cheaper today than using Indian-mined coal. And, of course, this advantage will grow as solar gets cheaper.

Of course, even in the sunniest parts of India, PV only produces electricity for an average of twelve hours a day. When people want light to read, cook or study, it isn't available.

(Solving this problem is what much of the rest of the book is about.) But one must note, too, that almost 20 per cent of all Indian electricity demand is used for pumping water for irrigation. This, as just one example, can easily be carried out solely in the daytime.

At the moment PV only provides a tiny fraction of Indian electricity. But it will grow rapidly with strong backing from the Modi government and from the favourable underlying economics. As in other countries around the world, it will then start to become increasingly costly to run the grid to cope with the unpredictability and diurnal variability of solar power. More PV means more batteries to help stabilise the voltage of the grid, for example in the event of unexpectedly high or low levels of sunshine. The cost of this needs to be factored in. And, very sensibly, KPMG also includes a cost for the financial impact of coal-fired power plants working fewer and fewer hours as solar soars. This is a real financial burden because, as in industrial countries, the fixed yearly costs of these power stations will be spread across a smaller electricity output.

What are the impacts of these charges, usually known as 'grid integration' costs, once PV has become a really significant portion of all electricity production? KPMG thinks the figure for India will be about 1.2 rupees (1.2 pence) per kilowatt hour in 2025. This is roughly in line with estimates for other countries. Even after including this figure, PV is still cheaper than coal by the middle of the next decade and then provides 12.5 per cent of all Indian electricity from about 166 gigawatts of installed capacity.

Most of the KPMG work is focused on the finances of building ground-mounted solar farms for large scale production. But it also looks at two specific applications: driving agricultural pumping operations (a task often performed by highly polluting diesel generating sets at the moment) and, second, what the accountants call the 'Solar House'. What do they mean by this?

> The concept of the 'Solar House' refers to the condition when the entire power needs of a household can be met by rooftop and on-site solar panels, which combined with energy storage can potentially make the household completely independent of the grid. This can happen when technology will bring the cost of solar power and storage systems to below the cost of power delivered by the grid. This event has the potential to change the dynamics of the power utility–customer relationship forever.

They go on to get really excited about Solar Houses. When have you ever seen accountants write like this before?

> The achievement of the 'Solar House' is expected to be a landmark in mankind's efforts to access energy. The 'Solar House' will help India leapfrog technologies in the area of supplying uninterrupted 24x7 energy to its citizens. When the conditions for the 'Solar House' are achieved, [it] can override all barriers.

KPMG expects 20 per cent of Indian houses to have PV by 2024/25. The authors make the point that once residential batteries have come down in price sufficiently, householders have a clear and unambiguous reason to switch to solar. It will be cheaper than being connected to what may be an unreliable and possibly expensive grid. And, as we may be seeing in other countries already, once households drift away from being connected, or even if they simply use far less electricity because of the PV on their roofs, the costs of the distribution network need to be spread ever more thickly on those that remain. That further increases the incentive on those people remaining on the grid to transition to PV.

Those who want the world to remain addicted to fossil fuels (and there are still surprisingly many of them) often use the argument that by denying the developing world the use of coal, renewable energy proponents are keeping billions of people impoverished. Coal is vital, they contend, for providing reasonably priced electricity in much of Africa, Asia and Latin America. There are several counter arguments, of which the most obvious is that many of the poorer parts of the world have no widespread electricity transmission networks, and therefore large centralised coal-fired power stations are an irrelevance. Coal-fired power stations do not, for example, help agricultural communities far from the electricity grid irrigate their crops. Solar power can.

India does have electricity networks across much of the country so coal-fired power might conceivably be helpful. But the second argument then comes into play: coal is no longer cheaper than PV across most of India. Solar is now the most cost effective, and least polluting, way of extending access to electricity to hundreds more millions of people. India continues to expand its coal-fired capacity – PV will take at least a couple of decades to completely wipe out fossil fuels – but as solar continues to fall in price a larger and larger fraction of new Indian generating capacity will come from PV. Poverty will be alleviated fastest by a continued strong commitment to solar, not from coal-fired stations that need a fifty year life to justify their investment costs. In a potent symbol of the inevitability of THE SWITCH, the world's largest coal mining company, Coal India Limited, decided in late 2015 to provide the capital to install up to a gigawatt of solar PV capacity across the country.

South American solar uptake

India and China's commitment to clean electricity is well known. Less noted are smaller countries that have made equally strong promises to encourage renewables, and PV in particular. October 2015 saw an auction in Chile in which solar projects convincingly undercut other sources of power. Global investment house Deutsche Bank put it as follows in a research note:

> Solar and wind are now cheaper sources of power generation in Chile than fossil fuels. According to results from the last round of tender in late October, renewables won 100 per cent of the tendered contracts for the supply of 1,200 gigawatt hours [of] energy ... Three solar parks offered to sell power between $65 and 68/megawatt hour, two wind farms bid a price of $79/megawatt hour and a solar thermal plant with storage offered power at $97/MWh. Meanwhile, coal power was offered for $85/megawatt hour in the same tender process.

The major problem facing solar developers in Chile is not the cost of the technology but the poor availability of transmission lines to take power from the northern regions of the country to the urban areas, including the capital Santiago. Chile's Atacama desert is one of the best places to collect the power of the sun in the world. However, it is over 1,000 km from Santiago and this will impede the growth of PV in the next few years. The country's 2015 installations, Deutsche Bank said, could exceed one gigawatt but this number will fall unless grid shortages are alleviated. Nevertheless, the commitment from the government appears unshakeable, with a promise that 45 per cent of the newly installed capacity between 2014 and 2025 would come from solar, wind and geothermal sources. In a country with substantial unmet energy needs and high electricity prices, solar will be the engine of growth in power production.

The same is true in Brazil. In December 2013, the price tendered by solar PV bidders for contracts to supply electricity was about $98 per megawatt hour. This fell to $89 in November 2014 and $78 a year later. In less than two years the tender offers have fallen by a full 20 per cent. And there is no shortage of bidders wanting to install solar farms in Brazil. Bloomberg asked some experts about the bidding process. 'Competition was huge in this auction,' said Rafael Brandao, a partner at Rio Alto Energia. '[Suppliers] were fighting for contracts and the prices were pressured at the end.'

Africa: mirroring the mobile phone

Nigeria, the most populous country in Africa with about 180 million inhabitants, has just begun to push solar. The existing electricity supply, even in major towns, is unreliable and blackouts are frequent. The government says that 96 million people do not have access to a power network. Some of these individuals are reliant on expensive and dirty diesel generators while many millions of others have no electricity of any form. Rather than push for greater amounts of large scale electricity production, the Federal government has decided to drive for a huge amount of new PV power, spread around the country to where people need it, particularly in the north. Vice President Osinbajo said in October 2015 that Nigerians would quickly witness 'a solar power revolution'.

He made those comments at a meeting in London. He was later asked by journalists there why the country didn't use its vast reserves of coal instead. His spoken reply, as transcribed by a Nigerian newspaper, is worth quoting in full.

We know that coal is available in some states like Kogi and Enugu but the main issue is that solar power, which entails the conversion of sunlight into electricity, is available everywhere. You don't need to transport it; you don't need to put it in a grid. It is much cheaper to buy some of the solar power equipment with about 200 dollars, which is about N50,000, and use it to provide electricity. The cost of using the grid is quite expensive and of course, very time-consuming. Consider the time you need to fix all the transmission lines and all of that; and we have 96 million Nigerians who have no access to electricity. If we want to deliver electricity quickly to that number of Nigerians at a relatively cheaper cost, we must take the solar route.

What comes across in all the reports of Nigeria's plans for solar is a profound sense that only solar can quickly provide power to the nearly 60 per cent of the country's population unconnected to its electricity networks. Politicians like Vice President Osinbajo seem to despair of making the existing generation and electricity distribution companies more effective. In a country with relatively weak central governmental machinery, solar allows the devolution of decision-making and fund-raising to local institutions, businesses and householders.

Many articles on solar power written by journalists in the developing world make a particular point with heartfelt intensity. To them, solar is like mobile phones whereas gas or coal power is analogous to the fixed line telephone networks. Many countries struggled for decades to push the fixed line telephone 'grid' out beyond the major cities. Usually, they failed. The costs were too high and expansion too difficult in countries with problems of graft and weak government institutions.

Along came the mobile phone and these difficulties largely disappeared. Nigeria is an excellent example. Today it has 150 million mobile subscribers compared to 200,000 active fixed lines. By contrast, in 2001 there were a quarter of a million mobile phones but 600,000 fixed lines. Governments and opinion formers instantly see the parallels between the failure of fixed line networks and the inability of large scale electricity grids to roll out services to the vast majority of the population. Both mobiles and solar PV can be bought and used by individuals or small businesses without having to work through inefficient large companies.

Microgrids and alleviating poverty

Writing with passion across Africa, journalists note the beneficial impacts of the arrival of almost universal mobile access on life in poorer countries and want the same thing to happen with solar electricity. Government ministers facing popular anger over supply disruptions and pollution from large power stations increasingly think the same. As with mobile phone networks, it is also possible to draw in foreign capital to pay for the installation of small scale 'microgrids' that employ solar PV as the core of their generation portfolio.

Microgrids – small networks linking a row of houses or an entire rural village – are bringing electricity to their first customers. Powerhive East Africa Ltd, the Kenyan subsidiary of a Californian company, now has the first licence held by a private company to generate and sell electricity to two counties in Western Kenya. According to the company:

> Powerhive was granted the concession as a result of more than two years successfully operating microgrid pilot projects powered by 100 per cent renewable energy in four villages in Kisii,

Kenya. The pilot projects serve approximately 1,500 people and have played a critical role in creating new businesses, enabling the use of productive appliances, powering schools, and displacing kerosene and diesel, which emit toxic pollutants.

Eighty kilowatts of power provide about 300 points of connection in Kisii. Other examples around the world show that tiny distribution networks using AC power (the sort you get in your home from the electricity network) can provide highly reliable and economical power to people who would otherwise have none. Almost all are based on increasingly cheap solar PV, combined with batteries and sometimes a backup generator. In countries like Kenya, where less than a quarter of the population has access to grid electricity, solar PV and microgrids will provide the basis of almost all growth in electricity generation.

Like Nigeria, Kenya has huge ambitions for the roll-out of electricity to its population. The country wants all its people, rural and urban, to have access to power by 2030. That's almost as fast as the expansion of mobile phone connectivity over the past decade, the obvious exemplar for this rate of growth. The target is simply not going to be achieved if Kenya tries to install large power stations and thousands of kilometres of high voltage power lines. But it might be if it uses hundreds of entrepreneurs to build local networks serving a few tens or hundreds of houses and businesses.

Critically, those entrepreneurs will rely on using a mobile phone payment system, bypassing the need for billing and cash collection. Kenya was, after all, the first country in the world to bring a mobile payments system into large scale use via the hugely successful M-Pesa technology and users will be able to 'top up' their electricity credit in the same way that they buy data for their phones. As with mobile, the country leapfrogged

Western payment systems, such as credit cards, with a next generation alternative much more suited to its own needs. It will probably do so with microgrids as well.

Zimbabwe also seems to have chosen solar. This country has an unreliable electricity grid with a working capacity of around 1,000 megawatts, less than 2 per cent of the UK equivalent. In October 2015 Zimbabwe and its partner China announced that Chinese money would finance the construction of 300 megawatts of solar by the end of 2017. The peak ouput will add thirty per cent to the amount of electricity available. The country has struggled with maintaining its existing fossil fuel power plants for decades. Solar will provide a more reliable daytime source of power. The solar farms will be constructed more quickly than could ever be the case for a coal-fired station. In a landlocked country like Zimbabwe with no pipelines, a gasfired power plant is an impossibility.

Even deeply troubled Burundi in central Africa is pushing ahead with PV. Political disturbances may delay the completion of a 7.5 megawatt park but the long-term future of solar is secure. Burundi is a country in which only one in twenty-five people has access to electricity and the country's total availability of power is about 52 megawatts. So this single new solar park will add almost 15 per cent to the generating capacity of the country.

The head of the development company, Yosef Abramovitz, told the *Guardian* in October 2015:

> Solar power is the engine of transformation in Africa. It can deploy so quickly that you can drive poverty alleviation and economic growth like no other technology. It's going to take off so fast now. Once you've shown the economic model, the ability to scale up quickly is there. We're going to see gigawatts of solar in the next five years.

Back in the USA: big company solar

At the other end of the development spectrum, the US is also embracing solar. Recent months have seen an increasing number of power purchase deals at remarkably low levels. Prices below $50 per megawatt hour have become almost unnoticed (although these are only possible because of a tax break). To get a sense of the importance of this, it may be helpful to know that SunShot, a US government-sponsored progamme, intended to get solar to near cost competitiveness with other forms of power by 2020. The target price was equivalent to $60 per megawatt hour for that year. The crop of surprisingly low prices recently offered has produced some scepticism. However, a recent paper by researchers at the government's Lawrence Berkeley laboratory maintained that $50/megawatt was 'for real', concluding:

> Although the U.S. utility-scale PV market is young and the operating history of many projects (particularly the record amount of new capacity built in 2013) is still very limited, a critical mass of project-level data now enables empirical analysis of this rapidly growing sector of the market. This paper draws on the increasing wealth of data to illuminate progress in PPA (power purchase agreements) prices, installed project prices, operating costs, and capacity factors. Using a pro-forma financial model, it also demonstrates that the recent remarkable decline of PPA prices to $50/MWh appears to be justified by the combined progress in installed prices and capacity factors.

In the last few pages I've included a large number of direct quotations from individuals, companies and government bodies working in the field of solar PV. I've done this partly to draw in comments from around the world but also to try to show that my thesis that PV is now the best form of power in which to invest is no longer just the opinion of wacky environmentalists. It is increasingly the view of the world's leading companies.

Among the most notable solar supporters is Apple, the world's most valuable company, which has made a promise of carbon neutrality for its operations and manufacturing plants around the world. As well as investing in solar in the US, the business has started to commission PV farms in China. One series of investments in Inner Mongolia will produce 170 megawatts of peak output across three huge farms. To get a sense of the scale of this, a solar portfolio of this size will produce enough electricity to meet about one thousandth of the UK's total annual power need. But to wipe out its entire carbon impact Apple will need to make investments of about ten times this size. For a company with $200bn of cash on its balance sheet, this represents little more than small change.

Apple is not alone. Google has made renewable investment pledges totalling $2.5 billion in recent years including financing the largest US residential PV installer, SolarCity. Global retailer IKEA has put 700,000 panels on its stores and recently pledged another €100 million to solar as it promised to move to 100 per cent clean energy by 2020. The world's largest retailer, Walmart, has made similar investment on its own warehouses while in the UK Sainsbury's has put over 100,000 panels on its roofs. Many companies have done similarly, some by investing in arrays on their roofs and others by committing to buy power from new solar farms. Unilever, the worldwide household products company, says it will buy all its electricity around the world from renewable sources by 2020 and a decade later expects to be selling its excess green power to the local communities in which it operates.

The value of investment by large companies such as these is three-fold. First, it demonstrates to the outside world that hard-headed businesses see PV as solid financial invest-ment. Second, the expertise of these companies in managing complex projects has helped make the whole sector more professional. Third, and perhaps most important, corporate purchasers of solar electricity usually make ten-, fifteen- or twenty-year commitments to buy power from the entrepreneurs seeking to finance their projects. This is of vital help in securing the investment the PV schemes need and ensuring that the developers can proceed with their plans.

To some people, growing corporate investment in solar (and in wind, which until recent months was getting more investment from big companies) is simply public relations. There is indeed probably an element of window-dressing in the actions of some businesses. They are seeking to appeal to environmentally conscious customers, and possibly investors. But most businesses are driven primarily by a desire to save money and avoid future electricity price rises. Discussing Apple's investments in renewable energy, CEO Tim Cook said recently that 'we expect to have very significant savings'. Similarly, Walmart – not usually seen as an environmental do-gooder – stresses the financial benefits of PV as much as its other benefits: 'Using the power of the sun and installing solar panels lowers our energy costs and is clearly good for the environment, but another benefit is that it keeps prices low for our customers.'

There it is. The largest retailer in the world, and a company that is famed for its ruthless and unsentimental approach to business, says that solar is already cheaper than the alternatives.

Chapter 3
New generation solar advances

Chris Case is the senior scientist at Oxford Photovoltaics, one of the world's leading solar PV start-ups. His job is to drive forward an entirely new way of making solar panels, using widely available and cheap materials known as 'perovskites', rather than the silicon that is conventionally employed. As with all solar PV technologies, the energy contained in photons of light pushes electrons in perovskite molecules out of their orbits around atoms. The electrons are then collected and turned into useful electric current flowing round a circuit. The use of perovskites promises to induce a further lurch downwards in the cost of solar power.

I had gone to interview Chris at his offices just outside Oxford as part of my research for this book. At the end of an instructive hour's conversation I noticed an old photograph on his desk. Clearly taken some years ago, it showed a house under construction, with solar panels in the process of being installed on the roof. 'Ah,' he said. 'That's a house I built in Providence, Rhode Island when I was finishing my PhD on photovoltaics in 1981. And I actually made those solar panels myself. I found out that a US manufacturer of PV cells was switching from making individual silicon cells that were four inches wide to the new global standard of ten centimetres. (Most conventional solar panels are made from many small cells wired together and assembled into what is known as a module. Cells are now much bigger than they were in 1981.) 'The company had surplus stock of the old size and I flew to California with some suitcases and bought enough to make about 3 kilowatts-worth of panels. When I got back to Rhode Island, I took all these four inch cells and assembled them into full-sized solar modules to go on my new roof.'

He leant across and showed me another photograph. In this one, he and a friend are actually making a solar panel. Chris is the person on the left, holding down a cell. I'm afraid the

How to make a solar panel in 1981. Chris Case is on the left. The beer cans were just the right size and weight to hold the cells in place.

obvious question immediately bubbled up. 'What are all the beer cans doing in your laboratory?' I asked, 'What you were doing with your colleague looks very precise work for people sharing more than a few drinks.' 'Well,' said Chris, 'a full beer can was exactly the right size and weight to hold down the cell so it adhered correctly to the glass on the front of the panel.' I moved back to more sensible questions.

Solar cells deteriorate if they are exposed to water. That's why they need to be encased in the dull glass that you see around today. When Chris had put all the small cells on the glass front surface and weighted them down temporarily with the cans, he connected them all up with wires and then put a glass waterproof cover on the back. He'd constructed some of the first ever home-made solar panels.

They had cost him a small fortune. He told me that the total bill for the 3 kilowatts of panels was about $10,000, which was about a third of the price he paid to construct the entire house. In today's money he had paid about $26,000 (around £17,000) to get solar electricity, even though he'd done the bulk of the work himself to make the full-sized modules.

I asked what had happened to the panels. Chris said he didn't own the house any more but had checked with the new owner a couple of years ago. The panels – now thirty-five years old – were working well, even though there had been some degradation in performance. He estimated that they were still producing 80 per cent of the electricity that they'd done in their first year. Importantly, every single one of the individual cells inside the modules was still working. The cans of beer had succeeded in the task given them.

Those 3 kilowatts of panels will have produced about 150 megawatt hours of electricity over the course of their life so far. The money saved has not yet come close to paying back the cost of the array. So it was a terrible investment for Chris if you assess things purely in terms of the cash value of the solar electricity generated. Nevertheless, his work in Rhode Island, alongside many other apparently eccentric pioneers, has helped make solar power cheaper for all of us.

Later in my research, I went back to the same science park just outside Oxford where Chris Case works. I parked my car in exactly the same place but walked to a different building, this time housing the laboratories of university researchers working on new methods of collecting solar energy. I was there to see Nina Klein, a doctoral student about the same age as Chris Case when he constructed his own solar panels.

I wanted to talk to Nina about the other main contender (alongside perovskites) to replace silicon. The research team she works in is interested in using simple carbon-based molecules that can also be engineered to generate a flow of electricity when exposed to light. It will eventually be cheaper, quicker and simpler than silicon to use these materials, Nina hopes. In fact, it will be possible literally to print them using machines not unlike newspaper printing presses. A light flexible plastic will be fed through the printing process and then collected on a reel at the other end.

Over the first eighteen months of her doctoral work, Nina built her own machine for making solar cells. This machine creates a near-perfect vacuum and she now uses it to produce very small solar cells from what are called 'organic' molecules. (In this context 'organic' means containing carbon.) Inside the vacuum chamber, a substance called pentacene (or one of many other simple carbon-based chemicals) is heated until it evaporates. The evaporated molecules disperse, unimpeded by air because of the vacuum, and are deposited on a glass layer coated with an electrical conductor on the top surface. The pentacene layer provides electrons when hit by photons of light. A very thin layer of a much larger carbon-based

Nina Klein working on her self-built vacuum deposition chamber

molecule called a fullerene then collects the electrons to create an electric current. By varying the temperature and length of time that the pentacene is allowed to evaporate, Nina can adjust the thickness of the deposit. Her aim is to create a layer just tens of nanometres thick. (There are 10 million nanometres in a centimetre – so this is a very thin smear indeed on the glass wafer). Her machine can use a type of stencil to create different patterns of deposition. This will enable her to run experiments that assess how to get the best electrical efficiency from a solar cell.

'Instead of building the machine myself, we could have tried to fund a purchase from a commercial supplier,' Nina explained. But it would have cost over £100,000, a sum that would be difficult to find from the limited grants available for solar research. Moreover, her direct experience of manufacturing the complex machine was helping her design her experiments. At present, her cells are coming out of the machine at about 0.5 per cent efficiency, a measure of how much of the energy in the light is converted to electricity. But she stresses that the point of solar PV research is not always to maximise electricity production. Her first experiments will combine the simple chemicals she uses to create a flow of electricity with quantum dots, tiny particles perhaps only 10 nanometres in size. She thinks that by creating hybrids like this that she'll understand new ways to achieve light conversion efficiencies that rival conventional silicon.

She's working on an entirely different approach to that used by Chris Case's team just a hundred metres away across the lawn. Many scientists in the field think that either perovskites or her simple organic molecules will dominate the collection of light energy for generating electricity in twenty years' time. The silicon solar panels of today may then be seen as an expensive and crude precursor to highly cost-effective solar energy produced by one or both of these two alternatives. Or it may even be that solar energy is collected by thin coatings on the surface of window glass.

Silicon is a very cheap raw material of limitless availability, needing little more than sand to make, while the manufacture of cells uses large amounts of energy and employs massively expensive machines. A single production line might cost $50 million. The research of people like Nina and Chris is needed to continue the decades-long process of continually reducing the cost of making solar panels.

Thirty-five years ago, Chris Case spent about $8,000 in today's money for each kilowatt of peak PV capacity he put on his roof. The cost would have been many times this amount if he'd used commercial solar panels rather than building them from a supplier's cast-off cells. Nina will probably see a smaller percentage change in the cost of solar PV in the next thirty-five years than Chris has done. But even that apparently reasonable view is less certain than we might think. It may be that in 2050 or thereabouts electricity from solar PV is close to free, much in the same way that computer storage capacity has reduced so much in price that it now would look almost costless to someone in 1981. One source calculates that the cost of the equipment needed to store one megabyte of data was about $1,000 in that year. By 2015 that figure was down to one three hundredth of a US cent.

A short history of solar panels

Science has known for almost 180 years that sunlight falling on certain types of molecules can generate a flow of electricity. Edmond Becquerel, a French scientist who went on to work on many different aspects of light, including early photography, showed in 1839 that certain metal compounds in acidic liquids produce a current when a portion is exposed to the sun. In 1883, the American Charles Fritts generated electricity by coating the element selenium with a thin layer of gold. This first solar cell converted about 1 per cent of the energy of light into electricity.

We can think of photons of light from the sun as nothing more than pulses of energy. Solar panels work by capturing this energy. Each pulse can dislodge an electron in the panel and give it the extra energy to cross a one-way junction between two thin layers of silicon (or other constituents). This creates a negative charge, adding to the electrical gradient between the two layers of silicon. If wires are attached to the back and front of the panel, the electron will flow back to the layer from which it originally came, creating useful electrical current.

This sounds simple, and in one sense it is. However, it took several decades before Einstein worked out that it was photons that imparted energy to electrons, causing the flow of electricity – the finding for which he won a Nobel Prize.

Russell Ohl, an American engineer at Bell Labs, developed a breakthrough understanding in the late 1930s of how to produce a more efficient cell using by creating a junction between two thin layers of silicon doped with tiny quantities of phosphorus or boron impurities. These dopants produce a natural surplus or deficit of electrons each side of the junction and the asymmetry is crucial to the creation of an efficient solar cell. Electrons that travel across the junction after capturing extra energy from photons of light produce an electric current when connected to a circuit.

Only in 1954, however, did US scientists make a working silicon-based solar cell that managed to turn more than 5 per cent of the energy in light into electricity. Making solar panels that convert larger and larger fractions of the energy into useful power at a low cost has been a difficult, long process since then. In one of the standard textbooks on solar cells, Professor Jenny Nelson explains how all solar cells work:

> Normally, when light is absorbed by matter, photons ... excite electrons to higher energy states within the material but the excited electrons quickly relax back to their ground state. In a photovoltaic device, however, there is some built-in asymmetry which pulls the excited electrons away before they can relax, and feeds them to an external circuit.

For those of us that are not scientists perhaps that is the only thing we need to remember about the technical aspects of PV.

Once knocked out of their usual orbit round an atom into a higher energy state, electrons must not be allowed to lapse back into their normal position around the atomic nucleus. Instead, we need to force them into doing some useful work by tricking them into crossing a junction through which they cannot return. They leave using an electric wire and provide us with a flow of electricity. The target of the hundreds of business and university departments working on solar materials is to find more efficient and cheaper ways of stopping electrons relax.

The scope for further technical improvement

Most of today's commercially built silicon-based panels still convert less than twenty per cent of the energy in the sun's light into electricity. This means that if 1,000 watts of light energy hits the surface of the panel a maximum of 200 watts of DC electric power can be generated. This is after sixty years of intense commercial development. The theoretical maximum efficiency from a panel with two silicon layers each doped with impurities, and a single junction between them, is about 34 per cent. Most scientists in the field think that the PV industry will struggle to raise the conversion efficiency of single junction silicon panels to more than 25 per cent. This is partly because only a portion of the spectrum of visible light – essentially the red end of the rainbow – has photons of the right energy to dislodge the outer electrons in silicon atoms.

However, if we use several different layers of semiconductor material, made from slightly different materials, we may be able to capture the energy from the blue end of the spectrum as well. This requires far more complex and much more expensive panels with several extra layers and multiple junctions between them. Research labs around the world are working to produce these multi-junction cells at reasonable prices. The maximum efficiency that will eventually be attained using several layers of silicon combined with other materials is probably about 50 per cent, but we don't know yet when this will be achieved or what the cost of the resulting panels will be. Nevertheless, that gives us the target; moving to between two and three times present-day efficiency, but at a cost that reduces the overall price per unit of electricity generated by a large percentage. And, of course, some other separate research is going into how to make the existing type of silicon cells in cheaper ways, without losing much efficiency.

I'll examine the scope for these cost improvements briefly before going on to the efficiency gains that silicon-based solar panels might see. Then I'll look at the radically new materials for making PV mentioned in the first pages of this chapter: perovskites and organic molecules.

Cost reduction

If we could make silicon for solar panels in ultra-thin slices, rather than in blocks, there'd be major benefits. At the moment the highly purified silicon required for solar panels is made into ingots using an energy intensive process at a very high temperature. The ingots are then left to cool and each block is sliced, rather like a loaf of bread, into individual very thin wafers. These small wafers of silicon, usually called cells, are then merged to form a full-sized module or panel. The process of sawing the ingots into wafers produces inevitable wastage in the form of silicon sawdust. One estimate is that half the pure silicon that has been expensively produced is wasted as dust. And, second, even after the most careful sawing, the thickness of the silicon is greater than it needs to be.

Several young companies around the world are looking to produce single wafers using completely different approaches that are almost as efficient as cells made from silicon ingots and which offer major advantages over conventional products; Solexel in California and NexWafe in Germany are good examples. Their initial focus is on improving the efficiency of the wafer at converting light to electricity.

Solexel will eventually become a full-scale manufacturer of panels, rather than simply trying to integrate its machines for making cells into the production lines of existing solar panel makers. The company claims its modules will be much lighter than the conventional silicon version, meaning that they will much easier to deploy, particularly on thin roofed buildings. In fact, the company claims, it will be cheap and convenient to fully integrate their cells into the structure of roofs, not simply to anchor them on the surface.

NexWafe is pushing towards an additional target of reducing the cost of making a standard complete PV module by 20 per cent as a result of cutting the bill for the silicon in half. Its first factory, scheduled for completion in late 2017, is also intended to reduce the amount of energy used in panel manufacture by almost 80 per cent. If successful – and not everybody is convinced that the approach of either company will work – the implications for PV costs are profound. In addition, the wider environmental impact of manufacturing techniques for ultra-pure silicon will be substantially lessened. (To make a PV module requires huge amounts of energy and the use of many toxic chemicals, albeit in small quantities.)

Another group of companies and researchers are trying to improve the percentage of light energy that is converted into electricity in standard sawn silicon cells. One promising technique uses what is called 'passivation' to improve efficiencies. This new process coats the rear side of the cell with a very fine layer of a material to stop excited electrons from recombining with atoms that are deficient in electrons. The leading Chinese manufacturer Trina Solar announced in December 2015 that its cells had reached a conversion efficiency of over 22 per cent using this technique. The best level of power collection for passivated cells, achieved in the laboratories of the University of New South Wales, the world leader in this area of scientific research, is about 25 per cent.

Eventual achievement of this level of performance in a standard silicon cell will push overall energy collection up by almost a quarter over levels currently achieved. Getting this increase in efficiency is not going to be expensive either in terms of materials or new capital equipment.

Other small efficiency improvements are in prospect, although single layer silicon solar cells are never likely to get much above 25 per cent in their conversion efficiency. But by improving mass-produced solar panels so that they achieve this enhanced output at the same time as decreasing silicon use by 50 per cent through the NexWafe or Solexel processes should result in the cost of full panels declining by at least 40 per cent. So, even if nothing else changes in the technology of collecting light and turning it into electricity, further substantial cost reductions will still take place.

Perovskites: Oxford Photovoltaics

Other prospective innovations radically change the nature of PV technology and may generate even larger efficiency gains. The most promsing looks to be the use of perovskites as the electricity generating material. Light shining on perovskite molecules dislodges electrons in the same way as with silicon.

We met Chris Case in the first pages of this chapter. He heads the technical team at Oxford Photovoltaics, a world-leading business attempting to commercialise the use of these molecules. The excitement around perovskite solar collectors arises because they are both cheap to make and have seen unprecedentedly rapid improvements in the efficiency of solar energy collection. In 2009 the best cells made of perovskites turned about 4 per cent of the sun's energy into electricity. That figure is now over 20 per cent, rivalling the best commercial silicon cells. The theoretical maximum efficiency of a single layer of perovskite is around 30 per cent.

Oxford Photovoltaics, as its name suggests, uses the fundamental research coming out of the university's solar research team. The leading scientist behind the new approach to solar power, Professor Henry Snaith, has rapidly accelerated perovskites to a position where they now attract global attention.

Perovskites are relatively simple molecules. Named after a mineral found in Russia's Ural Mountains in the nineteenth century, they all have a characteristic crystal structure. Oxford Photovoltaics uses a particular group of molecules called methyl ammonium lead halides. (A halide is an atom of a light gas like chlorine or bromine.) These molecules are good at capturing high energy photons in the blue portion of the spectrum unlike silicon's compatibility with reddish light. Slightly different perovskite molecules can collect the energy from varying portions of the solar spectrum.

Perovskites can be easily deposited as a very thin layer of liquid that rapidly dries on to a surface. Although silicon wafers are much thinner than they were, they are still 100 microns thick, or 0.1 of a millimetre, and they will not get much thinner. Perovskites generate electricity from far thinner layers than this. Eventually, Chris Case says, the energy world may be dominated by perovskite solar cells. Layers of subtly different perovskite molecules can be placed above

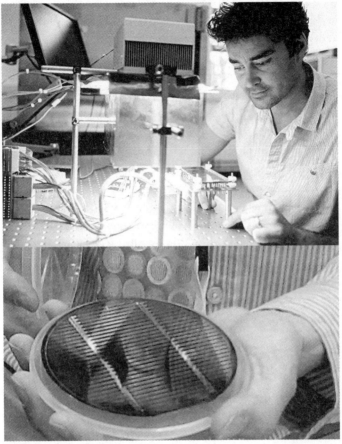

Professor Henry Snaith at work in his laboratory; and a tandem cell, with perovskites on top of silicon, produced by Oxford PV.

each other to build PV modules of the best possible efficiency. Each stratum of perovskites will collect light energy from distinct wavelengths and may push the percentage of total light energy captured up to above 40 per cent. However, the first ambition of Oxford PV is to get commercial manufacturers of solar panels to make what are called 'tandem' cells with a very thin layer of perovskite placed above a standard silicon cell.

The thin perovskite surface, married with layers of material above and below to collect the electricity, captures the energy from the blue part of the spectrum and the silicon cell takes energy from the red. The perovskite film blocks some of the light getting through to the silicon so the overall efficiency of the combined cell is less than the addition of the two separate layers. Put a perovskite collector on top of silicon and you may move the panel from 20 per cent to about 25 per cent efficiency, an increase of a quarter on today's best levels. This small change will rival all the prospective improvements that are likely ever to be made in single layer silicon cells.

The potential improvement is worth even more than you might at first think; it means an installer might be able to put panels with a maximum power of 5 kilowatts on the roof of a house that otherwise would only fit 4 kilowatts. The cost of the permitting, the labour, the scaffolding and the cabling will be almost the same. But these fixed costs will be spread over 25 per cent more electricity output. It's partly for this reason that the most efficient silicon panels today are priced at considerably higher levels per watt of output than the standard technology used in large solar farms placed on the ground and where space is not at a premium.

What's the cost of creating a tandem silicon and perovskite cell, I asked Chris. 'It probably adds about 5 per cent to the final bill for a module,' he replied. To get about a quarter more power for a cost increase of just 5 per cent is very good value, I commented, so why aren't manufacturers diving in today? 'Well,' he said, 'first we have to prove that the perovskite layer lasts. Our molecules are very sensitive to the presence of water, which will rapidly degrade its efficiency. So we've needed to prove that our properly encapsulated cells can survive long periods of bright sunlight and very high temperatures and humidity.' Case added that some buyers are also concerned about the presence of lead – correctly regarded as highly toxic – in the perovskite molecule. However, the amount of lead is tiny and is, of course, held permanently inside the glass layer that seals the solar module.

The second major obstacle is to persuade the big solar module manufacturers to add the capacity to coat silicon panels with a perovskite layer to the machinery on their existing production lines. Chris said that a silicon module production line involving twenty-eight different manufacturing processes costs about $50 million to install. The perovskite coating would need to happen in machines added to the processes after step 25 in the silicon manufacturing line. At that point there would be three perovskite-related machines and once the layer had been added then the module would be returned to the existing line.

A panel manufacturer needs to run its production line at full tilt twenty-four hours a day. Margins are tight in panel manufacturing and stopping the main production line even for a few weeks to install the perovskite equipment is risky and expensive. You can see why PV companies hesitate before making the commitment to the expense of the new machines, which Chris Case said would be between $5 and $10 million.

But, on the other hand, if the new coatings work, the manufacturer is successfully making panels worth at least a quarter, and perhaps as much as a half, more than conventional silicon. So the perovskite revolution will eventually happen because the financial arguments are so compelling. The first of the large scale PV producers to jump into this area will be taking a gamble on many things, including whether the new machines will work without disrupting the existing process and whether their customers believe that the new panels will last. However, if you run a PV manufacturing plant you also don't want to see your market disappearing in the direction of the people down the road who took the leap into perovskites first. So it's just a matter of time.

Once tandem perovskite/silicon panels are proved in the field (perhaps in a further three years or so?) we'll be bound to see manufacturers wanting to make multi-layer perovskite-only cells. They'll be cheaper than silicon and collect more energy than a tandem perovskite/silicon cell.

Oligomer cells: Heliatek

Not many European early stage businesses have raised €70 million in a single round of new finance. The German company Heliatek has managed it. Investment in new energy technologies has generally been disastrous for the funds that backed start-ups, and innovative companies pioneering new PV devices have been particularly bad investments. Its unusual success in raising capital is one of the many reasons why Heliatek is such a path-breaking company. This spin-out from the University of Dresden is the world leader in the commercialisation of an entirely new class of photovoltaic cells, based not on silicon or perovskites but on smallish carbon-based molecules called oligomers.

The underlying principle with oligomers is the same as for silicon; the energy of the photon of light dislodges an electron from its normal orbit around an atom and makes it available to be used as electric current. But from this point the differences grow. Silicon panels are heavy, rigid and require expensive metal supports in the field. They also need large amounts of energy. By contrast, Heliatek makes light, flexible and simple sheets of photovoltaic material. Organic (this means containing carbon atoms) photovoltaic material is 'printed' on a backing made from PET, the flexible packaging used for drinks bottles.

A Heliatek scientist holds the flexible PV film. The films weigh less than 500 grams per square metre and are less than 1 mm thick.

When I talked to Thomas Bickl, the head of sales and product development at Heliatek, he told me that their photovoltaic films require less than 10 per cent of the input energy of a silicon panel. This matters; a conventional panel can take two years to 'pay back' the energy used in its manufacture. A factory making Heliatek's films can also be 10 per cent of the size for the same volume of production. The films themselves weigh less than 5 per cent of a silicon module with a square metre weighing less than 500 grams. Of this less than 1 gram is the photovoltaic material itself. The rest is mostly the backing plastic and the cover that keeps the oligomers protected from oxygen and water, the dangerous enemies of all photovoltaic cells.

Nina Klein, the PhD researcher who we saw at the beginning of this chapter, uses a similar approach to making photovoltaics. Nina's machine creates a vacuum and heats the chemical until it evaporates, floats upwards and deposits itself as a very thin film on the backing panel. She is doing the painstaking research work by hand whereas Heliatek has a €10 million machine to do the job, passing the film through four separate processes to create a working device. Heliatek uses two different layers of active photovoltaic absorbers, each collecting a different part of the spectrum of the sun's light. These layers, and the electrodes that carry the dislodged electrons, are only 250 nanometres thick in total. (A sheet of paper is about 100,000 nanometres.) The whole film, including the front and back covers, is only 10 millimetres thick.

It is hugely important to the economics of this way of making photovoltaic cells that the process is 'reel to reel'. The virgin plastic film is mechanically unwound from a cylinder, passes through the four stages and is then stored on another reel at the end of the production line. The convenience of this is that the machinery is much less expensive, the whole process is automated and the finished cylinder can be shipped directly to customers, ready to be installed on a building.

Thomas Bickl went on to talk about the other advantages of organic photovoltaics. The manufacturing process is relatively simple compared to the large number of stages a conventional silicon panel goes through and it all occurs at lowish temperatures of 120 degrees or less. The film can be made in a large variety of widths to meet, for example, the needs of a particular building roof or wall. The walls of buildings are a natural site for Heliatek products and the company has installed its films as cladding on offices and factories. Architects are notoriously resistant to spoiling their beautiful constructions with solar panels but Heliatek films can add colour to otherwise dull surfaces. And the fact that the film is highly flexible and very light means it can be installed in many places where conventional panels would be inappropriate.

Heliatek film installed on a concrete building. The ultra-thin films can be mounted on glass or concrete to generate power.

'They're light, cheap to make, can be used almost anywhere, don't contain rare, expensive or toxic material,' I said, 'so why haven't organic photovoltaics taken over the PV market?' Bickl gave a candid reply. 'Silicon has had the advantage of half a century of development. It's become cheaper than anybody thought possible. Our materials will become cheaper per unit of electricity generated than silicon but only when we have entered very large scale production.' The first full-scale commercial factory that they build will make about 150 megawatts of panels a year, a tiny fraction of the world's need for photovoltaics, which might be as much as a thousand times that output by 2017.

Like Oxford Photovoltaics, Heliatek 'has to prove to the market that our film is as durable as silicon,' Bickl acknowledges. 'We need to show that after twenty-five years the product is still producing 80 per cent of the electricity it does when first installed. Our tests show we're not there yet. We're also not as efficient as silicon at the moment. The standard film currently converts about 7 per cent of the energy in light into electricity, although we can get much more in the laboratory.' Heliatek plans to get the efficiency of the film produced in factories up to 14 per cent at least within five years, still somewhat less than the 20 per cent currently achieved by silicon. (In February 2016, Heliatek announced that its best cells had reached a new record of over 13 per cent efficiency, at least in the laboratory.)

Against this current disadvantage in light capture, organic photovoltaics work much better in very hot conditions. Put a silicon panel in Saudi Arabia and its electricity output suffers badly. The productivity of Heliatek's film will actually improve. In addition, they work better than silicon in diffused light coming through clouds. So the comparison isn't straightforward because, in many circumstances, Heliatek's electricity output is much better than the single efficiency figure would indicate. The product's flexibility and light weight is also a significant attraction. Thomas believes that 'silicon will probably still be dominant in large systems where the heavy weight isn't an issue but we will be the best PV for use on existing surfaces such as roofs or façades'.

Heliatek gave me some figures for expectations of the cost of electricity generated from its film in 2020. A cloudy place like its hometown of Dresden might get electricity at around 11 euro cents per kilowatt hour (about 8.5 pence) while Saudi Arabia should be as low as 3 euro cents. To put this in context, in very hot countries with excellent sun, Heliatek is promising to supply power at about half the best cost of silicon PV today.

Organic photovoltaics, such as those from Heliatek, are more suited to being affixed to buildings than silicon. They're lighter, thinner and easier to wire up. When they are generating electricity, they are generally replacing expensive power that the building owner would otherwise have bought from the utility. This adds to the appeal. Electricity produced far from users in the middle of the Arizona desert will be very cheap, but power generated on the façade of an office building in Los Angeles is worth a lot more.

Like so many of my interviewees, Thomas Bickl went on to comment on the second crucial part of the transition to solar as the world's key energy source. We will move from large, centralised power stations – by which most commentators mean extensive wind and solar farms as well as fossil and nuclear plants – to a position in which most electricity is generated near to the point at which it is used. When the electricity consumer pays about two to three times what the distant electricity producer gets for the power, the economic logic behind this trend becomes overwhelming. Heliatek's films covering the south-facing façade of a standard house would provide most of the power a family needs, in Europe and pretty much everywhere around the world. There's usually far more unshaded space on the south facing façade of a house than on its roof making self-sufficiency easier to achieve.

Thomas Bickl also said that his sales team were continuously surprised by the inventiveness of the possible applications for Heliatek film that customers were ringing in about. Surfaces of all types, including the sunroofs of cars and the windows of buildings, can be used because the PV film is partly transparent to visible light. Car manufacturers, in particular, seemed to see huge potential in using a Heliatek surface on the roof of an electric car. In a sunny country a car parked outside might get much of its electric power directly from the sun rather than the mains.

I asked about the competition to Heliatek's organic PV films: 'Surely perovskites from companies like Oxford PV, layered on to the top of silicon cells, will offer much better solar collection levels than organic PV films?' Bickl agreed but said that 'conversion efficiency is not the only thing that matters – durability and cost are at least as important'.

The other main potential competitor Heliatek faces is a form of flexible panel made from another photovoltaic material called CIGS (Copper Indium Gallium Selenide). CIGS panels use well-understood technology but are much heavier than Heliatek's film. So they might work on roofs but are probably not a competitor on façades.

Improving solar: flattening the daily curve

This chapter has, so far, focused on the routes by which the amount of electricity generated by a single panel will be improved. But there are other ways of making PV more useful to us. Installers are concentrating on ensuring that solar power delivers a more stable flow of electricity over the course of the day and trying to avoid the peak for a few hours around midday as much as possible.

I'm writing this section of the book at half past seven on an October evening. At six minutes past seven my computer had received its usual estimate of the expected intensity of the sun tomorrow at each of about 150 points across the UK. I automatically put this information through a little program and my laptop forecasts how much each solar farm around the country will generate each hour tomorrow.

Meteo, the weather forecaster providing the forecast, thinks it's going to be a sunny day and my algorithm translates this into a sharp jump in electricity output. At 10am, all the solar farms and roofs in Britain will be producing about 1.9 gigawatts, rising to double this amount at 1pm and back down to around 1.9 gigawatts just after 4pm. This is not an ideal pattern for electricity suppliers or the operator of the national grid. When the UK has ten times as much solar power, this pattern will see output from PV panels rising and then falling far too fast for the grid operator to maintain the stability of the network. This is already a problem in places with a high penetration of solar PV, such as California. Even if electricity storage becomes very inexpensive, it would be much better to have a more constant pattern of generation.

In new PV farms around the world, project developers are finding ways of flattening this daily curve, trying to produce a more stable output during the course of the day. In countries with high late afternoon needs for electricity to drive air conditioning units, this is particularly important. There are two well-understood ways of creating a plateau of power across the entire solar day. They are usually called 'oversizing' and 'tracking'.

Oversizing

As panels become cheaper and cheaper, the overall cost of a large solar installation tends, of course, to fall. However, the prices of the other parts of a solar installation are probably not declining as fast as the silicon modules (although this is a conclusion not everybody agrees with). The most important, and most expensive, of the other components is the inverter, a box of electronics that takes the low voltage direct current from the panels and turns it into higher voltage alternating current that is precisely synchronised to the national electricity grid. Both domestic homes and large solar farms need inverters. In the typical house there might be one inverter but a multi-megawatt installation will have many, each one serving a particular part of a solar park. They usually sit in big metal boxes out in the field.

In the early days of PV, the installer always put in inverters of sufficient total capacity to cover the maximum output of the panels. If the midday power from the solar panels was 1 megawatt of direct current, the inverter capacity would be set to match that peak level. Of course, the maximum level would only ever be reached in the summer and then only for a few hours in the middle of day.

Cheaper and cheaper solar panels may make it sensible to use a different combination of panels and inverters. Typically the developer of a big solar farm in most parts of the world now wires up too many panels for the size of the inverter, a phenomenon called 'oversizing'. This means during the peak of a sunny day in summer the inverter would not be able to cope with the peak flow from all the panels. It deals with the problem of a potential excess supply by slightly detuning the panels and restricting the total electricity flow. The surplus power is lost as heat from the panels.

The effect of oversizing the solar array is to flatten the daily curve of power output, giving a much less peaky pattern of electricity delivery. In the chart below, based on the actual output of a 2.5 megawatt solar farm in south-western England on a clear day in mid-April, 5 megawatts of panels and a capped 2.2 megawatt inverter would deliver steady power for nine hours from 9am to 6pm. This is much easier for the operator of the electricity grid to deal with. (The peak of the

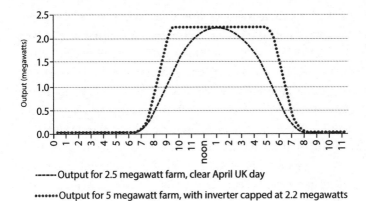

------ Output for 2.5 megawatt farm, clear April UK day

••••••• Output for 5 megawatt farm, with inverter capped at 2.2 megawatts

Building a solar farm with too many panels for the inverter to handle means a more even output throughout the day.

output of the conventional farm is at about 1pm, not noon, because of British Summer Time.)

As a result of oversizing, the expensive inverter, and the other items of kit in a PV system, are getting much better utilisation. For every thousand dollars spent on inverters, the owner gets more electricity output. In fact, the most advanced US solar farms can achieve average utilisation over the course of the year as high as 35 per cent. This is over three times the best levels of a country like the UK, where oversizing is almost unknown. As the sun is only over the horizon 50 per cent of the time a full year output of 35 per cent seems an almost implausibly high figure but the chart above shows how it might be possible in areas of very consistent sunshine.

As importantly, a larger fraction of the total power comes at the beginning and end of the day. In hot countries this is particularly helpful because peak national power demand tends to occur in late afternoon when air conditioning units are struggling to keep buildings cool but when the output from a conventional PV farm is starting to fall. The owners of solar assets also benefit because the price they get for electricity between 4 and 6pm can be very much higher than at midday.

Panels are likely to continue to fall as a percentage of the total cost of a PV installation and so more and more new solar farms around the world will be 'oversized'. Partly this is because it improves the returns from running the PV farm and partly because electricity utilities are going to put increasing pressure on solar developers to reduce the sharp peak in electricity output in the middle of the day. Flattening the curve of PV generation means that other generating sources don't have to ramp up and down their own electricity production as sharply. (A secondary effect will be that fossil fuel plants will need to work for fewer hours each day, further reducing the usefulness of new gas and coal power stations.)

Lastly, it is worth mentioning the importance of simply maximising the availability of electricity output. Whenever there is a bit of cloud, for example, we still want to meet the needs for power without turning on alternative sources. This may mean that eventually the world simply builds far too much PV capacity. The logic for this was forcibly put to me just after I'd just given a short presentation to an audience of executives from a large company, stressing the importance of energy storage to the solar future. Someone piped up from the side of the room. 'Actually,' he said, 'storage is not as important as people think it will be. Solar is going to become so cheap that the world should put far, far too much on the ground. This means that even on cloudy days, and in the early morning and late afternoon, there will be enough power to keep the electricity system going. Most of the time, this'll mean power is being wasted but this is easier to deal with than the consequences of a shortage.'

I recoiled slightly, perhaps because I'm old enough to be appalled by the idea of wasting electricity. But I realised that the questioner was completely right. Solar will make electricity eventually so cheap that it won't matter much if we over-install generating capacity in fields around the world in order to ensure that we always have enough during the daytime hours.

Tracking

The other way to flatten the daily curve of power output is to move the panel during the course of the day to follow the sun. This is called 'tracking' and can either be in one axis (up/down or left/right) or in both. Wherever you are in the world, a PV panel will produce most electricity when it is perpendicular to the sun so if tracking is sufficiently cheap it will always make sense to install it in a solar array. The greatest increment to electricity output occurs in countries closest

A 3-megawatt PV plant using dual axis trackers in Golmud, China

to the equator. One study in South Africa saw single tracking systems increase electricity generating by over 33 per cent. In Australia, a leading consumer website claims an extra 25 per cent for a solar array in Canberra. The benefits in the UK and further north would be much smaller.

At the beginning of the day, the panel is close to vertical because the sun is low in the sky. As the sun rises to its apex the panel is automatically flattened. And as the sun slips back towards the horizon each day, the position is shifted again to a more vertical angle. One of the advantages of using tilting panels is that the monthly variation in electricity output is far less than with a standard fixed installation.

Tracking isn't free. Not only does the machinery that tracks the elevation of the sun cost money but the amount of land area needed increases and the risk of mechanical failure of the tracking motors needs to be taken into account. One recent study suggested that a large scale tilting panel installation costs about 15 per cent more than a fixed position solar farm. However, almost all new US utility scale solar farms now use tracking systems, mostly of the up/down variety.

Cutting the costs of elements other than panels

Improving solar collection costs is important but reducing the bills for the electronics and wiring is just as critical. One study suggested in autumn 2015 that over three quarters of the final cost of a smallish residential PV system arose from what are known as the 'balance of system' elements such as the inverters that take the direct current output from the panels and turn it into alternating current finely coordinated with the grid.

These system components have declined in price some-what less than modules over the last few years. GTM

Research in the US recently said that PV module prices fell by 79 per cent between 2007 and 2015 whereas balance of system costs declined between 39–64 per cent depending on the type of installation and where it was being built.

As with the price of modules, there is still plenty of scope for reducing system costs. For residential installations, installers around the world can move to roof-mounting systems that allow modules to be anchored to individual mounts rather than long rails. This saves installation time as well as materials. Other innovations in immediate prospect include the use of higher power inverters. The impact on the total cost of the system will not be huge but the aggregate effect of many different small improvements will continue the decline in overall PV system cost. One recent study suggested that the slope of the experience curve for inverters may be as high as 18.9 per cent, a figure very similar to solar panels themselves.

Improving module efficiency, as Oxford PV proposes with its perovskite tandem cells, also helps. The labour cost of installing a thousand or a million panels will be the same, but the amount of electricity generated will perhaps be 25 per cent higher if we use better materials. The balance of system costs per dollar of revenue will be reduced in consequence.

Lastly, we are already seeing a switch to running large solar farms at a higher voltage in the DC portion of the installation. Many new arrays in the US now run at 1,500 volts, up from 1,000 volts (or 600 volt systems used prior to 2012). This makes the total cost of the system cheaper, even though some individual components are more expensive. One report estimated that this change alone will reduce the total cost of an installation by 3 per cent. As with many other improvements, the highly globalised nature of PV means that new technologies like this diffuse rapidly out from the early innovators.

Cheap new capital and levelised costs

Most media attention is directed at the cost of making solar modules. Sometimes we see articles on the price of other pieces of equipment a PV installation needs, but the number that really obsesses people is the cost of actually getting a solar panel shipped out of a factory in China or the US.

This is understandable. All of us are used to comparing the prices of physical objects and watching the changes. We listen to anxious voices on the news saying that poor harvests are pushing the price of bread up or the oil glut is causing the cost of petrol to fall. But in the case of PV, the price of the units is rivalled in importance by something very different: the cost of money. As PV gets ever cheaper, interest rates are becoming more and more crucial in determining the cost of electricity from solar farms. Nobody talks about this, but they should. It is now almost as important a consideration as the reductions in the cost of the panels themselves.

In the industry jargon the full 'levelised cost' of each unit of solar electricity rises sharply for each 1 per cent increase in the interest rate applied. This is true for panels on a small domestic roof and for a huge solar farm spread over several square kilo-metres. Perhaps the commercial installation required £100 million to build but it is virtually cost-free to operate. Its profit depends on only two things: the price it obtains for its electricity and the interest it pays on the capital used to build the farm. If we want to get developers to install hundreds of giga-watts of solar panels each year, we need to ensure the interest charges they pay are as low as possible.

The next chart shows why. It uses approximately accurate figures for the cost of installing and maintaining a large new UK solar farm and estimates the resulting cost of electricity and how this is affected by the cost of raising money. At a 3 per cent interest rate, the cost of the electricity coming from

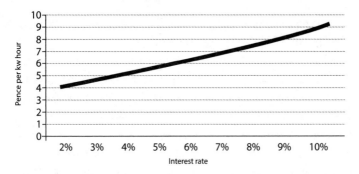

Levelised cost of electricity for a large UK solar farm, pence per kilowatt hour

the huge array is about 4.7 pence per kilowatt hour, rising to twice this if 10 per cent interest is applied. This is because a new PV installation has financial characteristics similar to the purchase of your house. You buy the expensive item once and then get the benefit of it for decades. Whether you can afford it or not depends on how much your bank charges to lend you the money. A conventional power station, such as one powered by gas, is different. It might have a construction cost of less than 10 per cent of a solar PV farm for each unit of expected output. It is therefore far less affected by the interest rate than PV. In contrast to solar photovoltaics, the finances of a gas-fired power station are much more vulnerable to changes in the price of its raw material, natural gas, than they are to interest rates.

I've suggested on earlier pages that doubling the accumulated number of solar panels ever made will cut the production cost of solar by about 20 per cent. If all other components of the PV system also fall in price by the same amount, then the underlying cost of power produced by the array will reduce by 20 per cent. We can get the same cut in the cost of electricity by reducing the interest rates demanded by banks or investors. A fall in the cost of financing a solar farm from 7 per cent to 5 per cent would have the same effect as a doubling of accumulated total PV panel production.

These two improvements are not mutually exclusive. Both can happen at the same time. And, to some extent, this is what has been going on. People have looked at the startlingly low prices offered by solar farm owners to electricity buyers and have assumed that all the reduction is coming from the falling cost of equipment. However, in recent years a large fraction of the decline has been driven by falling interest rates. (And a third consideration, which I'll discuss later, of improved assumptions about the useful life and reliability of solar panels.)

Ten years ago, when PV was almost unknown in countries other than Germany and Japan, lenders were reluctant to risk their money. As solar farms have spread across the world, investors no longer hold back. Electricity yields are predictable, operating costs have been seen to be reliably low and panel lives seem longer than expected. As a consequence, the interest rates demanded have fallen. As importantly, banks are prepared to lend a growing fraction of the total cost of the projects. Shareholders have to put in smaller percentages. This is good for overall financing costs because banks will generally demand lower interest payments than the dividends required by shareholders. (This is because they have first call on all the income coming in to the solar farm whereas shareholders can only get their dividends from the money left over after the banks have been paid.)

In situations where the required average interest rate on a PV project in 2005 might have been 10 per cent or more, we can now see figures as low as 2 per cent or so before inflation. These extremely low figures are usually associated with private installations on buildings or on community-owned solar farms. If, for example, you put PV on your roof there is really very little risk of not benefiting yourself from the investment.

It is as safe as the house. Similarly, as is the case in most of northern Europe, a cooperative of which you are a member rents a field outside the town and installs a small solar farm. The financial returns to these ventures do not need to be high.

The fit with the needs of pension funds

As well as improved familiarity, investors have also become increasingly aware of one of the attractive characteristics of solar farms. Their cash flows match pension payments. We shouldn't underestimate the importance of this around the world, not just in a few wealthy countries.

When a large PV station goes through the process of raising money, prior to being constructed, it will often seek an agreement from the local utility to buy all its electricity output for a set number of years, often more than a decade. This is called a 'Power Purchase Agreement' or PPA and usually fixes the price which is paid. (In places like Germany and the UK, the system is different and the government uses more complex methods to provide a secure price for the electricity.) This gives the financiers a good guarantee that they will get their money back from the sales of power. So pension funds can buy stakes in solar PV ventures and use the flows of regular cash from them to pay its clients a reasonably secure income.

Ten years ago, you might not have needed solar PV for your retirement income. Your pension company could have bought government bonds and remitted your monthly income using the returns from these ultra-safe investments. The interest rates were high enough to return a reasonable annual income from an individual's pot of pension money. That is no longer the case. As I write, UK and US government bonds that will mature in ten years' time would earn about 2 per cent interest each year, without any protection from inflation. A solar farm might be a much better investment, and in many places can provide inflation proofing.

Some pension funds have invested directly into large solar farms. In 2013, Lancashire County Council loaned £12 million to Westmill Solar, a member-owned cooperative venture which owns a five megawatt farm near Swindon that benefits from the UK government's Feed In Tariffs. The business pays an interest rate of 3.5 per cent on the money outstanding each year plus an amount equal to the rate of inflation. The loan will be repaid long before the panels stop generating. There's little risk to either the pension fund or to the coop's shareholders.

Mostly, large investors don't invest directly in projects like Westmill but put their money into funds which then in turn lend the cash to the biggest projects they can find. Alongside investments in other large and expensive assets that last for a long time and offer reliable cash flows, solar (and wind) are increasingly popular choices for investors. They are usually grouped together under the title 'infrastructure' which includes such other things as airports, tunnels and large hospitals.

One recent article in the magazine *Investment and Pensions Europe* summarised the reason for the trend:

> Powering the surge is a perfect storm of supportive policies and favourable financial and technological developments. Government policy aimed at stimulating investment in non-carbon power sources is paying off, while citizen groups are raising political pressure on lawmakers to move faster. Major investors are responding, raising allocations to an asset class that offers attractive yields in a low-interest-rate world, inflation protection from revenue often linked to inflation, and project finance structures that have more protections and control than rival asset classes, such as unsecured corporate debt.

The journalist Christopher O'Dea went on quote the comments of an industry leader:

> We're seeing the renewables sub-sector growing very, very significantly as a share of the overall infrastructure allocation at many clients,' says Andrew Robertson, senior managing director and co-head of Macquarie Infrastructure Debt Investment Solutions (MIDIS). Robertson estimates as much as 50 per cent of infrastructure loans being made are to renewables.

Governments cannot easily increase the rate of technological innovation in how solar panels are made (although they can create the buoyant market which makes driving down the costs of manufacture easier). However, good policy can definitely help drive down the cost of the finance that solar farm developers need. The simplest and most effective way of doing this is to guarantee the price to be paid for the electricity that is generated. That is why Lancashire Pension Fund was able to justify the apparently low rate of return it accepted at the Westmill Solar Farm in 2013.

Similarly, the huge boom in UK solar installations in the first quarter of 2015 was largely financed by institutional investors seeking a secure portfolio with a stable income protected from the effects of inflation. Although many of these investors waited until the asset had actually been constructed before providing their finance, those taking the initial risk were confident that their solar farms would be purchased by pension funds and insurance companies as soon as they were completed.

We shouldn't underestimate the scale of investment that might eventually be required. To provide from PV all the energy – not just electricity – that the UK currently uses, the country would have to put in place about 1.5 terawatts of PV.

That's almost 200 times what is on the ground and on roofs today. (We'll avoid some of this investment by also putting money behind complementary technologies such as wind power but we'll also have to invest in massive amounts of storage capacity.) At today's prices, just the solar PV would cost at least the total value of private pension funds in the UK, although solar will, of course, decline in cost.

If solar takes over as the dominant source of energy it will need to be both because of falling costs of equipment, and declining interest rates on the money that has to be borrowed. To get some feel for what is happening to the cost of money for solar projects I spoke to four people who organise fundraising for solar.

Funding solar

Gage Williams runs a renewables business on the western tip of England. His Cornish company has set up a variety of small wind and solar farms across the county, supplying power to users who usually have to pay higher than average rates for electricity because of Cornwall's distance from the main power stations. The German solar experts at the Fraunhofer Institute say that smaller scale companies like his can raise money more cheaply than the people who fund the biggest solar parks. Fraunhofer thinks that domestic and small commercial installations in Germany can be financed for at interest rates of about 2.4 per cent plus inflation. How does the UK compare?

At the moment West Country Renewables finances new projects, such as Gaia 11 kilowatt turbines for use on farms or 250 kilowatt solar parks, by raising about 60 per cent in loans from banks and around 40 per cent from shareholders. This is not unrepresentative for smaller scale developers. I asked Gage what he was paying for the money he has borrowed from financial institutions, in his case a local branch of the Swedish bank Handelsbanken in Bodmin, the pretty but isolated town in deepest Cornwall. He told me that his business was lent money at 3 per cent above the rate at which banks lend to each other. This inter-bank interest charge is usually called LIBOR and at the time of writing is about 0.5 per cent a year. At the moment, West Country Renewables is therefore able to borrow at about 3.5 per cent.

The rest of the finance comes from individual shareholders. It is difficult to accurately assess what these shareholders require as a return on their investment. If the business returned all its surplus funds to them then it might then pay about 8 per cent a year. This seems a reasonable level of required income. Merging the returns demanded by the bank and by the shareholders, one comes up with a figure of about 5.3 per cent for the average cost of capital. This is lower than most assessments of the cost for PV and for other renewables. But it seems about right for the sort of projects Gage is completing in Cornwall (although it is still considerably higher than the figures quoted by Fraunhofer).

I then spoke to Bill Weil, an investor in renewables with international private equity fund Ludgate, about the ways in which the cost of borrowing can fall. Bill is an American with substantial experience of investing in renewables around Europe and talks with a real belief in the inevitability of the move away from fossil fuels.

'It's important to understand the classes of money coming in to renewable energy,' he said at the beginning of our conversation. In the early years, it was cash from private equity funds. These funds invested in early solar projects, for example, because of the high returns possible because this was a new type of investment and risks were perceived to be high.

Gradually the investment community came to see that a large solar farm was actually a very secure investment indeed and what are called 'infrastructure funds' began to put their money into PV. Over the last few years the pattern has changed again and even conservative pension funds have started committing their cash.

'It wasn't long ago that pension funds would only invest once a solar project had been operational for a year or so,' said Bill. Then they decided that the performance of big PV farms was highly predictable and reliable so they bought in immediately on the completion of the scheme. Now, they'll stump up the funds to construct the farm once it has got planning permission and the developers have hired an experienced contractor for the building phase. It's taken a decade or so to get there but now the logical ultimate owners – funds providing a stream of income to pensioners – get involved when the big amounts of cash are actually needed.

As the sources of finance have changed, the costs have come down. Private equity, or 'venture capital' as it is sometimes known, required annual percentage returns in the mid-teens, said Bill Weil. The figure was around 8–12 per cent for the infrastructure funds. Pension funds are much lower still. Bill told me that the UK hasn't yet got down to the almost uniquely low German levels of required return. 'And that cheap financing, by the way,' he said, 'is why German solar costs are amongst the lowest in the world.' The crucial point is that solar PV is becoming increasingly attractive, everywhere around the world, to sources of finance that can live with relatively low, but highly reliable, rates of return.

The numbers I have just quoted are for the shareholder portion of the capital invested in solar. Some financiers will also use bank debt, just like Gage Williams, the investor in Cornish renewables I spoke to earlier. Pension funds, Bill commented, don't use outside bank financing. It actually might be more costly for a pension to borrow money from a bank than using its own money.

The best possible deals for all types of financiers are for schemes 'behind the meter'. That is, the PV is on a roof or in the immediate locality of an electricity user. All the electricity produced by the scheme is consumed by the building's occupiers. In these cases, the bill-payer is replacing electricity bought at high retail prices with his or her own generation at a lower cost. As we have seen already, this is far more financially advantageous than supplying to the local utility, which will only pay wholesale rates for the power produced, which are usually less than half retail prices.

In Germany, which puts substantial levies on electricity delivered to homes and smaller businesses, the difference is even starker. In 2015 a domestic user paid about 22 pence per kilowatt hour while wholesale electricity usually trade below 4 pence, creating a five-fold difference in price. Financiers are understandably keener to put their money into PV that replaces retail electricity because they know that commercial solar farms are exposed to the long decline in wholesale electricity prices taking place in Germany and elsewhere.

PV placed on domestic houses is 'behind the meter' and the UK Department of Energy and Climate Change said in December 2015 that the cost of money put into home solar installations is about 4.8 per cent in inflation adjusted and pre-tax terms. That is the rate of return a householder typically demands. This suggests a substantial recent decline; a leading economics consulting firm had suggested in 2011 that the figure was somewhere between 6 and 9 per cent, and would stay at this level at least until 2020.

Bill Weil gave his opinion that most subsidies for solar will disappear over the next year or two around the world.

'Governments will be clumsy about it, just as they have been in Britain and elsewhere.' The rapid disappearance of feed-in tariffs and other credits will shock the finance sector, which may temporarily withdraw support for solar. 'Capital doesn't like change,' he said, 'and there are lots of other types of investment to put money behind'. However, the investment strike will eventually pass. After the 'behind the meter' projects which don't need subsidy, the first category to return will be backing for extensive solar farms with committed investment grade customers, such as the large private companies like WalMart and Unilever.

Many large companies, of which Apple is the most visible today, want to buy a guaranteed supply of low carbon electricity and therefore sign contracts with solar farms for all their output. A credit-worthy customer committing to buy the electricity output provides banks with the security that they need to finance the largest solar farms.

Bill went on to talk about the long-term drift of intensive energy users to geographical places where power will be cheap. Solar will eventually provide electricity at what would seem today remarkably low prices. If you run a business that smelts aluminium – a process that uses staggering amounts of electricity – you will move your smelters to areas where solar or other renewables are available in bulk and at low cost. 'Aluminium,' he said, 'is basically stored energy.' Its manufacture will use up much of any temporary surpluses of low carbon electricity that the world will generate in a solar future.

He mentioned a precursor of this future shift. Silicor, a major manufacturer of PV-grade pure silicon, is opening a large plant in Iceland because of its need for electricity. Iceland has abundant supplies of deep geothermal heat that can be used to drive steam turbines to make electricity. When solar is fully developed, this example will be copied across industries and continents. Energy intensive activities, such as metals smelting and chemicals manufacture, will be concentrated in places like North Africa, Australia or the south west of the US. This global shift is entirely predictable and will help speed up the transition to an entirely renewables-based energy system, Bill Weil concludes.

Richard Nourse is head of Greencoat Capital, a fund manager in London specialising in renewable energy assets, including solar PV. He told me that money invested in large solar farms of ten or twenty times the size that Gage Williams is financing in Cornwall is requiring returns of about 6.5 per cent, including inflation. That's more than it costs Gage's business. Money is particularly available for projects for which there is some form of government guarantee for the price of the electricity produced. He expects the required rate of return will probably go down further as investors get ever more confident about solar power.

Although the returns to solar investment, which are spread out over several decades, are near-perfect for pension funds, he says that he still finds it difficult to pull in money from these investors. 'It's still not an asset that they understand,' he said to me in a slightly exasperated tone of voice. Pension fund managers have difficulties owning portions of physical assets which require some degree of involvement in management. They are used to simply holding shares in companies quoted on stock exchanges which can be bought and sold in seconds. If it buys shares in Vodafone or Glaxo, a fund has no responsibilities and no obligations. Buying a solar farm is very different, even if outside professional engineers are hired to ensure that it is operated well.

In line with what Richard Nourse had said, one pension fund manager indicated to me that her funds would be uncomfortable taking a stake of less than £80 million in any single working wind or solar enterprise. That makes traditional 5 or 10 megawatt PV sites, which cost less than £1 million per megawatt to construct, far too small to be attractive to invest in.

For the solar revolution to succeed around the world, this will have to change. The huge amounts of cash seeking stable long-term returns need to be provided with appropriate investment vehicles that match their liabilities but also do not require them to manage physical assets. In addition, I think it absolutely necessary for pension funds to reduce their management costs so that they can hold smaller assets economically.

The pattern that I'd discerned in my interviews thus far was clear, although a little dispiriting. Developers are getting better and better deals from sources of finance in the UK and elsewhere but the rates of interest demanded aren't yet as low as they might be. Then in late December 2015 I got a call from Peter Sermol, a financier who works for a small investment bank in London. Peter told me about the terms he'd just agreed with major pension fund providers to put capital into a highly innovative new scheme.

A large UK city's housing division wanted PV on the 5,500 homes in its portfolio with east, south and west facing roofs. The benefit to tenants would come from lower electricity bills because they'd be able to use the solar electricity at no cost. The council needed a PV provider to agree to install solar panels on its social homes in return for keeping the subsidy payments (feed in tariffs) from the electricity generated. The UK government had just announced a deep cut in the subsidy for home solar and most companies had backed away because they couldn't raise the money to put the PV on roofs. But Peter's sources of capital, which include some of the biggest pension funds in the UK, were prepared to provide his company with funding at rates of about 2 per cent plus the rate of inflation.

This highly competitive rate, plus some very low prices from the company installing the panels, meant that the homes will get their solar panels. This is the lowest cost of finance I've seen in the UK – it matches what is available in Germany – and it is only possible because the government's feed in tariffs, although now much lower than they were, still rise every year with inflation. Providers find this highly attractive because it matches their obligation to increase the pensions they pay by the rate of inflation each year. A city council is a very good credit risk and therefore Peter was able to attract money into the scheme at an unprecedented rate of interest. The result is that he can finance PV on domestic roofs providing social tenants with free solar electricity at a cost that is less than 7 pence per kilowatt hour. The low cost of finance means that PV on these domestic roofs is now almost at 'grid parity' with UK wholesale electricity prices.

That example is an unusual case. A very high quality borrower putting reliable PV on a large group of houses represents an asset that delivers a near-guaranteed inflation-linked return to pension funds. This may be a model for the future but is the money available around the world for others to copy this scheme or, at the other extreme, those of entrepreneurs like Gage Williams whose projects should also be financed at low rates?

How much capital is available for The Switch?

I then rang Andy Moylan at Preqin to ask a simple question. How much capital is available around the world to finance the construction of solar farms? Preqin is the largest provider of data on what is termed 'alternative assets', which include such things as venture capital funds and investments in solar. If anybody knows it would be Andy.

Andy told me that Preqin tracks the amount of money going into funds that specialise in financing infrastructure, including solar farms. The worldwide amount of money going into the funds that seek to invest in these assets has been running at about $160 billion a year. Of this the amount going into solar energy has been $15–20 billion a year over the recent period, or around ten per cent of the total. (The comments of Andrew Robertson of MIDIS quoted on p. 103 suggest that this percentage may have risen sharply in recent months.)

Very roughly, about $80 billion was spent globally on installing solar PV in 2015, so infrastructure funds provided about a fifth of the capital required. Other sources included governments and utilities as well as the bank balances of homeowners who had paid to put arrays on their roofs. To continue to grow at the 40 per cent per year of recent decades, solar needs to be able to pull in more and more capital from institutional investors for the next few decades. Andy said that Preqin's research suggested that managers of large pools of capital, such as pension funds, were showing increasing interest in infrastructure investments such as PV.

A survey that Preqin had just carried out showed that almost half the fund managers they'd interviewed wanted to increase the allocation of funds to infrastructure. Their target for the percentage of their funds that they wanted to invest in this type of asset has risen from 4.9 per cent to 6.3 per cent in the last four years, an increase of almost 30 per cent. Solar may form only a fraction of this total, which also has to pay for investments in water supply in growing cities, the construction of electricity grids and the development of public housing, but PV is an increasingly visible class of asset to the fund managers who allocate investment.

Large amounts of capital are piling up in pension funds and insurance companies around the world. But could the funds available conceivably be sufficient to pay for up to $3 trillion that will be needed to be invested each year by about 2035? Three trillion dollars is twenty times the money put into infrastructure funds over each of the last few years. It seems an impossibly difficult target but Andy dropped me an email after our interview to say that Preqin's estimate of the total pool of capital available to finance infrastructure assets around the world was about £80 trillion. In other words, once solar becomes fully accepted as an 'investable' asset – that is, something that institutions believe is appropriate for their portfolios – then the amount of money available is definitely sufficient to fund THE SWITCH. The growth of PV is not going to be held up by a shortage of capital. The more difficult question is, of course, whether solar will give high enough returns to be attractive to these asset managers.

Overall, the pattern I discerned from these conversations was that solar PV was becoming an increasingly attractive asset for banks to lend to and investors to buy. In some specific situations, the returns demanded are already at rock-bottom levels of around 2 per cent, whether for German community PV farms or panels on council-owned homes in a British city. For countries or companies making strategic investments, the returns may be similarly low. But for what might be called 'mainstream' deals in which large investors take substantial stakes in solar farms producing tens of megawatts of power, financiers require higher dividends or interest payments. However, these rates are also falling from levels of around 8 per cent to 6 per cent or lower. This alone might cut the cost of electricity from solar panels by over 20 per cent. If this decline is to continue, then the crucial requirement is for stable government policy and a high degree of confidence among investors and bankers that PV farms will be able to sell their electricity at reliable prices.

The lifetime of solar panels

The third and final contribution to cost reductions after technology improvements and falling financing costs is the extension in the lifetime of panels. A panel that works for longer produces electricity at a lower cost than one whose output falls away with age. The same amount of initial cost is spread over a longer life, and therefore greater electricity production.

As little as ten years ago, it was assumed that the electricity output of panels would decrease by about 1 per cent a year from the date of installation. One rule of thumb is that panels should be replaced once their output falls to less than eighty per cent of the initial level. This meant that the expected working life of an array would be about twenty years if output fell in a straight line. The actual performance decline is now thought to be typically 0.3 per cent a year, or possibly even less. Degradation from ultraviolet light is less and the electrical connections are better. At 0.3 per cent annual output reduction, the panel wouldn't be replaced for more than sixty years if the owners followed the usual rules. Nevertheless, the working assumption now is that solar panels will be replaced after about thirty-five years.

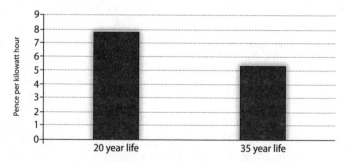

The reduction in cost for PV if the panels last for thirty-five years rather than twenty years is around twenty per cent.

I'll compare two lengths of panel life – twenty and thirty-five years. If we take a conventional solar farm, built in the UK at current prices of around £0.8 million per megawatt, financed at a five per cent rate of interest, then the impact of extending the life of a panel from twenty years to thirty-five years is to reduce the implied cost of electricity by 20 per cent. This percentage difference would be even greater if interest rates were lower.

There's still some scepticism about panels lasting thirty-five years or more. But industry-leading manufacturers, such as Trina Solar, now guarantee that each of their panels will produce more than 80 per cent of their initial output at the end of the twenty-fifth year of operation, promising a rate of decline of no more than 0.7 per cent per annum over the entire life of the panel. The average performance across its millions of modules will be better, meaning that thirty-five years is increasingly seen as the correct figure to use when approaching financiers for cash to develop a solar farm.

Alongside improving technology and lower costs of funds, extensions of the life of solar farms are decreasing the cost of solar power around the world. Of course, the move to a world energy system based around solar energy will not be smooth but all the forces affecting PV are improving its position. As its costs fall and performance improves, sources of capital will regard investment in PV as increasingly attractive. And this in turn will make investors more and more willing to fund the huge factories that will help commercialise the advances of companies like Oxford Photovoltaics and Heliatek.

Chapter 4
When the sun doesn't shine

PV is going to be extremely low cost, will deliver power in sunny places fairly evenly throughout the day and will work without human intervention for several decades. But it will need to be complemented by other energy collection technologies, and by various types of energy storage (see Chapters 6 and 7). This chapter focuses on the other renewables that we can expect to see employed in significant arrays to help provide a more consistent supply of electricity than is generated by PV alone. In some countries these alternatives will actually be more important than PV. For example, Morocco has made a commitment to what is known as 'concentrating solar power' and many countries in high latitudes will use cheap wind technologies for low-cost power, Denmark being a good example.

Although PV looks as though it is going to be the dominant energy collection technology worldwide – and many tropical countries will need very little to supplement it other than overnight storage – substantial supporting roles will be played by the other options in this chapter.

Concentrating solar power

Ten years ago, before the recent dive in solar PV prices, it looked as though the best way to capture the sun's energy would be to employ 'concentrating solar power' plants (CSP). The big advantage of this technology is that it can store solar power as heat, as well as produce electricity during the day.

CSP comes in several very different forms. In many of the early ventures, long rows of semi-circular mirrors focused the sun's light on to a thin tube containing a liquid. This liquid heats up to temperatures of several hundred degrees and flows to a generating hall where it is allowed to evaporate into steam. The steam then drives turbines to produce electricity. The big plus is that the heat can be held for several hours in a storage tank and used to produce power when the sun is down. It works best in very hot and sunny countries and the approach wouldn't be feasible in the UK, for example.

Spain actively backed this approach to power generation for several years. The first example of a large scale working plant was Andasol, sited in the province of Granada in the south of the country. A 150-megawatt plant, it produces more power than a comparably sized PV farm and can continue generating electricity for almost eight hours in the dark. However, it failed to start a trend, and CSP appeared to be losing appeal. Its cost has not fallen at remotely the same rate as photovoltaics and – until the last couple of years – interest had waned.

Other engineering approaches have been tried to see if they can make the finances more attractive. In the most common of these today, circles of individual mirrors all reflect light on to the apex of a tall central tower. This structure captures the sun's energy from thousands of mirrors and uses it to heat a fluid. The super-heated liquid is then conveyed to the bottom of the tower where it is evaporated and used to drive a turbine.

The innovative Crescent Dunes CSP plant in Nevada.

Perhaps the most striking recent example of the use of these heliostats, as these mirrors are properly known, and a single collecting tower is at the Crescent Dunes project in Nevada, US. There, 110 megawatts of power can be generated at peak production. Its total productivity is about twice that of a PV plant of similar rated capacity. Moreover, the heat it collects can be stored for up to ten hours, meaning it can keep producing some power for almost the entire twenty-four-hour cycle. However, with over 10,000 heliostats, it was expensive. The bill for construction was almost four times the price of photovoltaics that would have produced the same amount of electricity.

Despite the high cost, CSP is starting to gain interest again in countries with high solar radiance. We're seeing that high temperature heat storage, and the consequent ability to provide power during the hours of darkness, is very valuable. SolarReserve, the builders of the Crescent Dunes plant, are also constructing CSP facilities in South Africa and Chile. They will be providing either guaranteed power across the entire night or focusing their delivery of electricity into periods when the grid is under stress and prices are high. In South Africa, more power is needed in the early morning and late afternoon and the main utilities pay much more for solar power produced at these times. As the company's CEO pointed out in an interview: 'The ability to shift the delivery of the electricity to when [the grid operators] want it is worth about 2.6 times the underlying rate in South Africa.' The other innovation from SolarReserve is combining the CSP plant with a large amount of PV on the same site; the photovoltaics deliver during the day and the CSP will take over to provide electricity at night and in the early morning.

The most visible commitment to concentrating solar power, however, currently comes from Morocco, where the government is planning several huge plants that – combined with PV, hydro and wind – will make 50 per cent of the country's electricity renewable by 2020. In a reversal of the recent trend towards larger and larger solar towers sitting at the centre of circles of mirrors, Morocco has initially gone back to rows of parabolic troughs at the world's single largest CSP project. The Ouarzazate complex of four huge plants will be five times the size of Crescent Dunes and supply enough power for around a million homes. At around $9 billion, the cost is eye-wateringly high although this facility will provide highly reliable power in a country that currently imports virtually all its existing sources of energy.

The long-term ambition of Morocco is to send surplus power from its renewable sources to other countries, including those in Europe. The country sits immediately to the south of Spain and a 500-km power line should enable it to export solar electricity to its northern neighbour. This would be an expensive route to build but pylon lines of this length are now almost routine in places such as Brazil and China. THE SWITCH will be much easier to achieve if the world is able to move surplus power around more easily than at present.

Helio100

Perhaps the world will choose some mixture of CSP and photovoltaics in tropical countries. It may be possible to pull down the cost of CSP to levels that make it comparable with PV. Certainly there's no unavoidable piece of science that makes it impossible to achieve improvements. One contender for much less expensive CSP is a new South African technology based on significantly cheaper components. As with many other small scale renewable technologies, innovative young companies are challenging the assumption that the lowest cost of power will come from the biggest possible size and complexity of installation. Helio100, a spin-out from Stellenbosch University, is not attempting to build the most efficient or largest system for energy collection but aiming for the lowest cost, most widely usable solution with cheap heliostat stands holding mirrors that automatically track the sun 'plonked' down in a field. (That's their word, not mine, by the way.)

I've been writing about new energy technologies for the best part of a decade and one thing I know with certainty is that businesses are always too optimistic about the eventual

The Helio100 trial installation in South Africa.

cost of their innovation. Nevertheless, I think it is worth saying that the Helio100 developers suggest an eventual cost per kilowatt hour in a very sunny country of less than 5 pence, including the cost of full-scale storage for night use.

If Helio100 turns out to be right in its estimates, it will be cheaper than any new fossil fuel plant anywhere in the world. The cost is probably less than half what larger, utility scale, CSP plants can achieve, and, notably, cheaper than PV combined with electricity storage in batteries in any part of the world. So, let's be cynical and say the Helio100 people are almost certainly being too optimistic and double their figures. That makes 10 pence per kilowatt hour, which is still highly competitive with diesel generators or any other form of highly decentralised power generation that has a significant amount of storage.

Simple heliostats combined with heat storage may turn out to be the best way of providing 24-hour electricity in some remote parts of the world.

Wind

Of the 90,000 terawatts of solar energy hitting the planet's surface at this moment about 1 per cent is transformed into wind. The bulk of this energy moves from the tropics to the temperate zones, taking heat away from the hotter regions. However, there are also more localised breezes that move air from seas to coastlines and up hills, and these tend to be the easiest to capture in wind farms.

The amount of energy in the wind, across the whole of the globe including the oceans, is about 900 terawatts. (Remember that the planet will need about 30 terawatts to give all ten billion people a continuous supply of 3 kilowatts of power in 2035 so wind is still easily sufficient, at least in theory, to meet world requirements by itself.)

Whether wind is cheaper than solar PV today depends entirely on location. In the middle of a land-locked tropical desert, electricity from wind will cost several times energy from PV. On the west coast of Scotland, though, the reverse is true. In the best locations with easy access to transmission lines, wind is probably now cheaper in the UK than any other conceivable means of generating power, and the same is probably true everywhere on coastlines around the North Sea and in other high wind regions such as parts of the Great Plains of the US.

Richard Nourse (who I spoke to about investment costs in Chapter 2) told me that he thought the cost of onshore wind power in good UK locations was now well below £70 per megawatt hour. A new gas-fired power station would struggle to achieve that level – although a fully depreciated elderly coal-fired plant might be cheaper – and the wind figure is probably at least 25 per cent below the price a government would have to offer to get another new nuclear power station built. Nevertheless, the cost of wind in Britain should actually be much lower than it is. Despite excellent wind speeds in coastal areas, the UK is an expensive place to put turbines. Land rentals, planning costs and grid connection charges are high compared to many other places. The country also blocks the newest, biggest and cheapest types of wind turbines being constructed onshore.

In the US a recent estimate by a respected government agency put the average figure for the cost of electricity from new wind developments at around $66 a megawatt hour, compared to projections of well over $70 for gas, even with very cheap fuel prices. Wind costs have fallen 15 per cent since the same survey of projects completed three years earlier. The $66 figure assumes a cost of capital of almost 9 per cent. Take that figure down to 6 per cent, a more reasonable figure in a world of very low interest rates, and the cost declines to less than $55. In the best locations, such as parts of Texas, the number would be below $50. In places like this, wind can still be cheaper than solar.

So why should we assume PV is a better long-term solution, even in most of the world's windy locations? The reason is that photovoltaics are on a much steeper experience curve, and even if they are not cheaper now in some locations, they will almost certainly become so within a few years. The International Energy Agency, a body that remains deeply suspicious about the growth of solar, sees PV falling in price by up to 45 per cent between 2014 and 2040, while wind only falls by 30 per cent.

Most estimates for the rate of decline of wind costs set a figure of about 10 per cent for each doubling of accumulated 'experience', expressed as the total production capacity of wind turbines ever built. Wind turbine farms will continue to be able to provide electricity at lower and lower prices but because of the shallower slope of the experience curve and because PV is growing much more rapidly and so gaining improved costs, solar will eventually be much cheaper. In only a few benighted places will wind eventually end up the better alternative from the point of view of cost. Even in the UK solar will eventually beat large wind turbines.

Wind turbines are also less good at collecting energy per unit of space employed. Turbines might capture about 2 watts of energy per square metre, while PV in a good location harnesses five times this amount. So to meet a nation's needs, more area needs to be given over to wind (although, unlike solar, agriculture can still take place over almost all the land area underneath a farm of turbines).

Even if the underlying cost of electricity produced from wind is higher than PV, there may well be good reason for a society to bear that increased burden. In high latitude locations, winds tend to be stronger in the periods of the year when the sun is low in the sky. Wind complements PV very well although, unlike solar energy, the predictability of its monthly output is low.

Wind turbines aren't popular everywhere, although only in the UK is there a concerted campaign to block all new installations across the whole country. The number of suitable locations in which large onshore wind farms can be installed is not unlimited but the UK could comfortably accept far more were it not for substantial local opposition. Germany has about three times the number of turbines per unit of land area, and Denmark has more. However, big new farms need good grid connections with the capacity to accept substantial volumes of electricity and these locations are not in unlimited supply.

Two trends follow on from the declining acceptability of large scale onshore wind in some countries. (The lack of space certainly isn't a worldwide problem.) First, developers are moving to offshore locations and second, there's another burst of entrepreneurial energy aimed at producing really low cost smaller wind turbines which can be more easily fitted in to overloaded grids.

Wind blows more reliably offshore and so the average outputs of turbines are much higher. The productivity of wind farms is often measured as the percentage of the maximum output achieved over a period of time. (Wind turbines don't always produce more and more power as the wind speed increases. Output rises sharply until the wind is blowing at about 25 miles an hour. Above this level electricity generation plateaus until very high speeds are reached and the turbine is then automatically shut down to protect its moving parts.)

The average UK offshore wind turbine achieved 37 per cent of its maximum production in 2014, compared to just over 26 per cent for onshore farms. One way to better complement PV output is to put more turbines offshore and to make them bigger and sited as far away from the coast as possible to catch the highest possible winds. However, this is much more expensive than onshore wind. Despite the higher speeds, electricity from offshore wind is currently about twice as costly as the best onshore locations and although costs are coming down sharply, it doesn't look as though offshore wind is ever going to catch up with PV, even in areas with little sun.

Many options are being pursued for getting lower costs offshore – mostly by making things bigger. The central idea is that a monster turbine isn't twice the price of one half its size. I'm unconvinced by this approach and think that more radical solutions are needed. The idea that looks most intriguing to me is the Hywind floating turbine, a machine that is anchored to the sea bed. The first commercial installation, financed and developed by the Norwegian energy giant Statoil, will be fifteen miles of the coast near Peterhead, just north of Aberdeen. The turbine itself is a standard 6-megawatt turbine from

Siemens and the difference is that the tower is not sunk into foundations on the sea bed. Instead it floats, anchored in place by three cables stretching out to 60-tonne concrete blocks to hold it in place.

The first farm at Peterhead will be in water at least ninety metres deep, twice the depth usually regarded as the limit for standard offshore turbines. Five huge structures will sit in a four square kilometre farm. The cost of this first venture will certainly not be any cheaper than comparable fixed towers, but the windspeeds will be exceptional and the farm will be generating by 2017, says Statoil. The Hywind turbines are quicker and easier to install, a crucial advantage in seas where several weeks can pass before conditions are calm enough to carry out conventional offshore installations.

Spinetic

The second conept for wind is even more revolutionary: that instead of focusing on making larger, taller and more powerful turbines, it might be better to design the cheapest, easiest to install wind collectors that the world has ever seen.

As I mentioned when discussing Helio100 a few pages ago, we're seeing a move to small scale, highly decentralised energy converters across different technologies. PV on domestic roofs combined with a few kilowatt hours storage in a battery is just the most obvious example. Put PV or wind in a large field miles from anywhere and your turbine or solar farm will operate more efficiently but you will get a fraction of the financial benefit you would if it was in your garden. As governments around the world remove renewables subsidies as fast as they can without too much embarrassment, the financial incentives to move away from large scale industrial renewables grow. Even the most anti-renewables government will find it difficult to stop ordinary people putting electricity generation 'behind their meter'.

That's where Spinetic sees its opportunity. I first came across this little company after speaking at a university panel event on sustainability. When you work in this field, people often come up to you after public discussions with what might politely be described as eccentric ideas for new approaches to low carbon energy. Many of them breach the laws of thermodynamics. On this occasion, the pitch was for a small vertical wind turbine. Most turbines of the conventional three-bladed type are called horizontal because the drive shaft that passes through the turbine is parallel to the ground. These wind turbines, being created by a small company called Spinetic, had blades that rotated around a vertical axis.

Vertical turbines have been a nightmare for investors. Although the engineering suggests that they could be simpler and more reliable than conventional turbines, the first commercial designs have all been catastrophic technical and financial failures. 'Why should Spinetic's design be any different?' I sceptically inquired of the surprisingly normal looking person who'd approached me. 'The designers come from the automotive component industry,' was the apparently irrelevant reply.

The one thing I know about car components is that they are built to extremely high standards of reliability and almost ridiculously low cost. When a single seatbelt in a single Tesla recently malfunctioned, every car of that type had to be recalled at enormous expense. Component manufacturers supplying to the car industry live every day in the knowledge that a single failure could bankrupt them with costs and legal bills. And despite the requirement for almost impossible levels of product reliability, outside manufacturers are also told to make things for pennies that in any other industry would cost pounds. Your car may have seemed expensive when you paid for it but it is an utterly remarkable machine for the price.

Knowing the designers of the Spinetic turbine had spent years in an environment of brutal efficiency drives combined with almost paranoid fear of product failure made me very interested and a few weeks later I visited an otherwise unremarkable farm off the road between Oxford and Swindon and saw a little turbine installation whirring away on the top of some scaffolding. It had been there for several windy winter months and, unlike almost any other vertical wind turbine anywhere after that length of time, was still working.

Over the next few months Spinetic negotiated terms for its first round of investment. With a million pounds or so in the bank (including a very small amount from me), its people hunkered down in a small warehouse outside Swindon for three years, barely communicating with the outside world. New versions of the turbine appeared at infrequent intervals and were tried in the fields of the farm I'd visited. If successful, the trial versions were shipped down to the windswept Cornish fields of one of the other shareholders where they were exposed to more gales than land-locked Wiltshire gets. The engineers were not just working out how to make a turbine that survived intense storms but also how to make all the components as cheaply as possible. How thin can the vertical aluminium blades be and still not buckle in 100-mile-an-hour winds? How little copper could they get away with in the alternator that generates the electricity? Could all the main components other than the blades be made using 3D-printing or injection-moulding of plastics?

Like many engineers before them, the designers were desperate to keep the details of the turbine secret from the eyes of potential competitors. Photographs were banned, lest they should give any clue to the remarkable innovations

A Spinetic wind 'fence' installed in a secret location in Wiltshire, UK.

that Spinetic had introduced. Patent application after patent application was made to provide some protection. Finally, in December 2015, after almost six years of design work, the final prototype was approved and installed for testing on a farm not too far from the factory. The turbines will sit in panels of five, about four metres from the ground, ready to feed electricity into an inverter that will link them to a home, or a dairy farm's milking parlour perhaps, a battery or a small office. Panels of turbines can be tied together in long rows across the middle of fields. With supports dug into the ground using a farmer's own machinery, this is a product that will work almost everywhere around the world where wind speeds are high enough.

All this is very interesting, but what about the numbers? Spinetic is cagey, even with its shareholders. However, the talk is of being able to generate electricity at prices below 5 pence a kilowatt hour, a number usually thought of as about the average price of wholesale power over the last few years. That's less than half the price a domestic user pays after transmission bills and various other costs. This strikingly low figure will be achieved, they say, on sites with average wind speeds of more than five metres a second (just over ten miles an hour) at near ground level. That means you won't be able to put the Spinetic turbine at the bottom of your urban garden and make it pay. But much of the open land on the Atlantic coast of the UK meets the requirement, as do many other west-facing coasts around the world and large parts of South America and East Africa.

If it works as planned, this machine will be the first small turbine in the world to deliver power at costs equivalent to the cheapest new fossil fuel plants. It will require no subsidy for farmers and landowners in Britain or around windy parts of the world. As with big wind turbines, it will complement the output of PV arrays, meaning that a user seeking to reduce purchases of grid electricity will probably want to own a solar array as well for the low-wind months.

Spinetic's approach mimics Helio100 in South Africa in using a 'plonkable' technology requiring little skill to install. Even more important, the components can be made and then assembled in countries without sophisticated manufacturing plants: 3D printers and machines for bending aluminium will be all that is required.

Who knows whether these vertical turbines will deliver on the figures above? The record of vertical turbines is so dismal that no rational person would expect success. But another, much more conventional wind turbine is also showing that small machines may be able to compete against the big commercial wind farms. Britwind, a resurrection of a bankrupt existing UK manufacturer, has launched a 15-kilowatt turbine of the right size for a small farm. Standing about 25 metres high, the turbine maker promises to reduce the cost of the electricity it produces to around 60 per cent of the level of the most competitive existing smaller turbines. On a good site that would mean it would make financial sense without any subsidy, as long as the owner used all the electricity themselves.

Biomass

The world's gas, oil and coal reserves are the final outcome of the decay of living plants and small organisms hundreds of millions of years ago. Instead of using these fuels for our energy, we can utilise plant matter that has only recently grown. In fact, the only source of energy for many people in poorer countries without electricity is the wood or charcoal that they use for cooking. In other countries, such as the UK, biomass is burnt in power stations that formerly used coal as their fuel. Each year Drax power station in Yorkshire burns about 4 million tonnes of pellets made from wood and straw to generate about 4 per cent of Britain's electricity.

Alternatively, we can allow plant wastes to rot inside huge sealed tanks called anaerobic digesters (AD), which exclude air. Over about six weeks, this organic matter decays, producing water, some solids and a large amount of what is known as 'biogas'. The principal component of biogas is methane, the main constituent of the conventional gas which is piped to our homes for use in central heating boilers. This gas is usually burnt in turbines which produce electric power.

Using biomass for electricity generation can be a helpful complement to solar power. However, plants and trees aren't very good at using their leaves to collect the energy of the sun. Typical examples such as an ear of corn or the leaves of an oak tree only capture and retain only 1 or 2 per cent of the energy they receive every day for use in the photosynthesis process. By contrast, of course, a solar panel manages nearly 20 per cent. However, it cannot hold the power it generates even for a second, whereas the typical plant is a perfect device for storing energy. That's why we eat them, of course, to capture both their taste and their energy value. A food calorie is a unit of energy.

Across the globe, plants and the photosynthetic organisms in the water, such as algae and cyanobacteria, transform the sun into chemical energy at a rate of about 90 terawatts. That's about 0.1 per cent of the amount of the sun hitting the earth's surface. Although an individual plant in a well-watered and fertile location might successfully convert over 1 per cent of the energy it is receiving, large areas of the world are too dry or too wet to get full value from solar radiation. In other places the soil is too thin to support full-scale plant growth. Seventy per cent of the earth's surface is ocean, where photosynthesis occurs in bacteria and algae, but in which the process is generally slow and even more inefficient.

In theory, if we need 30 terawatts to give everybody on the planet a continuous stream of 3 kilowatts, then the 90 terawatts potentially provided by the products of photosynthesis might seem sufficient to meet the world's needs. However, burning wood in an old coal-fired power station such as Drax is a highly inefficient way of converting the energy because the ratio of electricity to the energy value of what is burnt is at best about 40 per cent. Other newer technologies such as anaerobic digestion or gasification cannot do much better. Nevertheless we could, at least in theory, get a substantial fraction of the world's total energy need from biomass, either by burning it for electricity or by turning it into combustible gas for heating.

But should we do this? Most experts are strongly opposed to increased use of plant matter and food for energy. The serious risk is that humankind would simply do what the Easter Islanders appear to have done, chopping down all the trees on the island for fuel and for other purposes, leaving a barren, unproductive landscape. Rich tropical rainforests are particularly good at capturing solar energy through photo-synthesis. But if we cut trees down, and do not replace them, not only will we eventually run out of fuel but we will also have reduced the planet's ability to store the carbon dioxide produced by the burning of fossil fuels. One of the side ben-efits of photosynthesis, of course, is 'carbon capture' as plants consume the CO_2 to help build their branches and leaves.

We don't have to make the same mistake as on Easter Island; we can carefully manage plantations so that we sow plants and trees to replace the amount cut down. We could start to farm the forests of the world on an unprecedented scale. It might be sustainable, at least in theory.

Virtually nobody wants to do this. Our global experience has been that trying to use the products of photosynthesis for energy causes serious problems. We need plants for calories, both for us and our food animals. Using biomass for energy alongside our needs for food and for CO_2 sequestration involves a conflict. Every hectare that is used for bioenergy may be a hectare lost to food production.

However, it may be worth pointing out that we don't actually need a large fraction of the 90 terawatts of photosynthesis-generated energy for food. If we gave all the ten billion people in the world in 2050 a diet of 2,500 calories a day from plant matter, we'd need about 1 terawatt of food energy, or approximately 1 per cent of all the energy captured from the sun by photosynthesis. (As an aside, you might notice at this point that our energy requirements from food are little more than one thirtieth of humankind's overall needs for electricity and other supplies of power.)

However, the minimum need for us to utilise 1 terawatt of energy for food tells only part of the story. Although vegetarians are a substantial fraction of the population in some countries, such as in many parts of India, most cultures choose to get many of their calories from meat and dairy products. A cow or a chicken is far from a perfect converter of plant matter to meat calories – it might take eight calories of grain or grass to produce one calorie of beef. So realistically the world doesn't need just 1 terawatt of food energy, it probably needs 2 or even 3 terawatts if we include the food eaten by animals that will eventually become meat on our tables. But even 3 terawatts is still a very long way short of the ninety terawatts available from photosynthesis. We should not completely rule out using plants and trees for some of our non-solar energy. In particular, it may be possible to use some portion of the world's land to grow much more biomass than at present, and therefore not diverting food or forests.

We may also be able to find ways of adding to the 90 terawatts of energy capture by the process of photosynthesis. And do so in ways that increase, rather than decrease, the availability of food calories for human consumption. Tropical Power hopes to do this.

Tropical Power

Large parts of the world's land surface are simply too dry for reliable food production from photosynthesis. Additionally, and this is much more of a problem than simple aridity, large areas of the globe have good rainfall that is concentrated in just a few weeks of the year. This is true, for example, of a large part of Africa. Seeds of crops such as wheat or sorghum may sprout and the food plant begin to develop, but then the drought hits and the crop eventually perishes.

Tropical Power in Kenya is working on getting plants to grow on land that is currently semi-desert in the north of Kenya. The plants will then rot in an aerobic digester, and the methane that is generated will be burnt to produce electricity. The huge advantage of anaerobic digestion is that the methane can be stored for hours, or even a few days, until needed for electricity generation. It is a near-perfect complement to PV in countries with good daily sun.

What sort of plants grow well on arid land that might only receive one or two periods of rainfall a year? Tropical Power has set up trial plantations of two types of crop on land in Laikipia County, north of Nairobi, that would otherwise only support small amounts of scrubby vegetation. The plants – *Opuntia ficus-indica*, better known as Prickly Pear, and *Euphorbia Tirucalli*, often called Pencil Cactus – are part of the CAM (Crassulacean Acid Metabolism) family, which contains thousands of different species, including succulent agaves, cacti and pineapples. CAM plants conserve water by shutting the pores on their leaves during the day. This stops most of the evaporation that would otherwise happen. At night the pores, or stomata, open and the plant takes in CO_2. It then temporarily stores the gas using a chemical reaction that employs the carbon dioxide to make a simple acid.

When the sun returns the following day, the acid gives up the CO_2 and it becomes available to start the photosynthesis process that makes the complex carbon-based molecules that the plant needs to grow. This two stage process of capturing the CO_2 during the cooler hours of the night and carrying out photosynthesis when the sun's light is available is very different to conventional crops. Most plants both take in CO_2 and use it for photosynthesis during the solar day.

CAM plants evolved to cope with highly variable water availability. Not only do they close the pores on their leaves

Young Prickly Pears on the Kenyan plantation.

during the heat of the day but many species are also able to store large volumes of water in their leaves. Over the course of a year, they may require only about a tenth of the water needed by a conventional food crop such as a maize plant of the same weight.

Many CAM plants appear to be able to grow well in areas that are too arid for crops or trees. Tropical Power says it will be possible to grow at least ten and possibly twenty tonnes of dried biomass per hectare each year. This is more than can be harvested from willow coppicing (another biomass crop) of a similar-sized area in Britain, where the rain falls every month.

What will Tropical Power do with the CAM plants that it harvests? It cannot burn them to drive a steam turbine and make electricity because of the high moisture content of the leaves. But these plants are broken down well in anaerobic digesters by bacteria and other micro-organisms that thrive in the absence of oxygen. CAM plants contain relatively little indigestible lignin (the woody material in trees and shrubs), further improving the yield of gas from the digester. Anaerobic digestion produces a gas that contains CO_2 and methane as well as trace amounts of other compounds.

The biogas is piped out of the digester and can be stored and then burnt in a gas turbine that generates electricity. One academic study showed that Pencil Cactus can produce over 100 litres of methane for each kilogramme of dry plant matter. When the methane is burnt, it turns into water and carbon dioxide. Nevertheless, this electric power can properly be called 'carbon neutral'. All the CO_2 from the biogas and from the burning of the methane is derived from carbon dioxide initially captured by the CAM plant.

Most AD plants around the world burn their biogas at a constant rate in order to deliver a stable and reliable stream of electricity to the local grid. Until recently, that was what national electricity authorities wanted, so the gas turbines at AD plants are appropriately sized to meet this consistent level of output. Tropical Power is different. Its power plant has a much greater capacity than the average amount of gas coming out of the digestion process would suggest.

I asked Mike Mason, the chairman of Tropical Power, about this. He said that the crucial reason was that their AD plant was intended to act as a mirror to a local PV farm that his company is developing. The Kenyan electricity grid did not want another large PV supply coming on to their system, supplying huge amounts of electricity during the day, but nothing after 6pm. Mike and his colleagues reasoned that if their AD plant could store the gas coming off the rotting plants during the day, it could burn it at night to produce a level stream of power. He therefore oversized the gas turbines that burn the methane, enabling Tropical Power to make larger amounts of electricity for ten or twelve hours a day, rather than a smaller but constant amount over the whole twenty-four hours.

Why, I commented to Mike, had the world not already thought about growing CAM plants on dry lands? He thinks the reason is that (pineapples excepted) few CAM species have any worth as foodstuffs. Scientists have known for decades of the high energy content of some CAM plants but until recently this was thought to have no financial value. As anaerobic digestion has begun to get more effective, and the value of plant biomass as a means of storing energy for periods when the sun isn't blazing becomes more obvious, interest will grow.

The Kenyan field trials will work out which particular CAM species will work best, growing quickly and reliably, and also proving easy to harvest. A successful energy crop must be mechanically harvestable using standard agricultural equipment, and should simply regrow once it has been cut rather than having to be re-sown. And it should be palatable to pastoral animals, otherwise there's a risk that the species will sprout like a weed, eventually invading other land that is used for food crops. But if sheep and goats like the taste, it won't spread far outside the harvested area.

Mason's research hasn't yet told him which species meet all his requirements but he's optimistic about Prickly Pears. He showed me some photographs of the trials fields his company has planted in Kenya. After ten months, the plants are well established and look as though they are thriving. Tropical Power has just commissioned its first AD plant near where the feedstock is growing. (In fact, Mike thinks this might be the only one in sub-Saharan Africa.) The complex range of micro-organisms that eat the complex molecules in plants take some time to settle in so when I interview him he doesn't know how well the digester will eventually work.

He's keen to stress the associated advantages of digesting CAM plants in dry areas. First, he says, the water stored in leaves has a very high value in itself. It might be possible to drain off some of the nutrient rich water from the AD domes and use it to grow a separate crop of free-floating aquatic plants. These are called Lemnacae and they are particularly attractive because they grow extremely rapidly in the right conditions, doubling in size within a day. They can contain over 25 per cent of their mass as protein, and are therefore valuable as feed for cattle or for fish. Mike is eager to see whether it will be possible for small farmers to develop a business farming tilapia, an increasingly popular fish around the world, using these plants for their food.

And, second, not to be ignored, is the value of the digestate, the semi-solid material that comes out of the AD plant at the end of the digestion activity. This has considerable potential value as a soil-improver. So while the area in which the digester is located might be too arid for crops, it might be possible to combine growing CAM plants with a small amount of conventional agriculture, aided by some water, nutrients and the digestate from the AD plant.

Mike is also working on improving the design of anaerobic digestion. When I first heard him lecturing about his projects he was speaking lyrically about how much better cows are at dissolving green matter than a modern digester costing several million pounds to build. 'A cow, or indeed a kangaroo, can do in 24 hours something that takes a huge steel dome six weeks to do,' he complained to an audience of academic researchers. 'Why can't anaerobic digesters just copy how Daisy does it?'

The Laikipia anaerobic digester.

This is a man who is prepared to get his knowledge from detailed hands-on research. His talk continued with a description of working with cows that have had a large tap inserted in one of their stomachs at birth. Researchers can open the tap and out flows some of the contents of the cow's digestive system. (He assures me that the animal feels nothing and continues eating placidly as this process goes on.) Some of his work looks at the chemical and mechanical activity going on in the stomach and mouth. The cow's digestion works rapidly, he hypothesises, because it repeatedly chews the leaves and grass in the diet, smearing each tiny piece of grass with a saliva that assists the microbes in breaking down the tough cellulose molecules to get at the energy they contain. Mike proposes a new design for much smaller and cheaper AD plants that will be suitable for individual farms across the world, copying some aspects of what cows do. Simple to operate and easy to maintain, he believes that this will help small farms develop an important source of secondary income from electricity sales. 'To have global appeal,' he said to me, 'anaerobic digestion needs to make financial sense using very small digesters. Otherwise micro-farmers will end up shipping their harvested CAM plant leaves long distances, reducing the incentive to grow these new crops and probably ruining the financial attractiveness of the anaerobic digester.'

When I interviewed Mike Mason, I wanted him to focus exclusively on these plans for growing biomass on arid land and generating electricity from better AD plants without reducing food production. But it wasn't possible to get him to stick to the topic. Mike also spoke excitedly of another innovation he is piloting at Laikipia. Using photographs from satellites, he can plot the passage of clouds across his part of Kenya during the solar day. The amount of power coming from an individual solar farm is severely affected by the degree of shading. If a bank of cloud is five miles to the west of his solar installation, moving at ten miles an hour, his software can predict the fall in output of the farm in thirty minutes time. This gives plenty of time to adjust the gas turbine attached to the AD plant to start producing more power as the shading starts. This is of vital importance to the stability of the Kenyan grid and if his iPhone app works, it will help the managers of solar parks around the world better integrate their installations into a stable electricity grid.

As is usual with serial entrepreneurs like Mason, he doesn't even stop there. His mind is already turning on to another series of opportunities in Kenya. He recently approached the government for permission to site his next large solar farm alongside a hydro-electric dam. Most dams around the world produce power intermittently. Over the course of the year, a dam might only be working at full capacity a small portion of the time. Mike says that it's logical to build PV farms next to these dams. They should be run to hold back water during the day, perhaps operating at 25 per cent of their capacity. The PV farm will provide the electricity instead during the daylight hours. As night falls, the dam's output will be increased.

One of the crucial advantages of his proposal is that Kenya doesn't need to invest in a new transmission grid for evacuating the electricity from the PV farm. It's already there to service the hydro-electric dam. However today the transmission line rarely takes electricity out at full capacity. Adding the PV should mean both that electricity supply is more stable and that the pylons and substations are better utilised. Often renewables are criticised because they are supposed to be expensive to integrate into the existing electricity grid. This is one of many cases when the reverse will be true.

The government has told Mason he can try out his proposal on a smallish 60-megawatt dam. (I say 'smallish' but 60 megawatts is over two per cent of Kenya's total generating capacity as of 2015.) If it works, he can go on to the main hydro-electric plants on the Tana river, which are ten times as big and provide a significant fraction of Kenya's power generation capacity. If he can make his idea work on these dams, he'll have cracked the problem and entrepreneurs around the world can try to do the same thing at other locations.

The global potential for CAM plants

Tropical Power and its academic partners, including Kathy Willis, the head of science at Kew Botanic Gardens, have started to make estimates of how much of the world's energy might be provided by CAM plants grown on arid lands.

They've worked out a figure for the amount of land that is too dry to support conventional crops but which does have enough rainfall to support water-conserving CAM species. The group bases its estimate on numbers from the Food and

Agriculture Organization (FAO), the Rome-based research arm of the UN. The FAO define 'semi-arid' as land with rainfall of between 300–800 mm (12–31 inches) if rains fall in summer and 200–500 mm (8–20 inches) if they come in winter.

These figures highlight the impact of getting rains in just one, or possibly two, periods a year. If the rainfall in some parts of eastern England all just came in one season lasting a couple of months, the land would be described as semi-arid. Cambridge, for example, gets about 560 mm (22 inches) a year and yet it is within a few miles of some of the most productive wheatlands in the world. Agriculture is possible there because the rain that does fall is spread throughout the growing season, and evaporation is relatively unimportant.

The FAO says that about 12 per cent of the world's land is semi-arid. A much wider area is left unused for arable agriculture because of poor productivity due to lack of water. In total, almost 20 per cent of the land surface might be available for growing CAM plants without a measurable impact on food production. Tropical Power uses a slightly more conservative figure of 2.5 billion hectares, about one sixth of all land on the planet.

Not all of this land can be converted to (indirect) energy production in the way that Tropical Power is doing in Kenya. But much can. How much energy could these 2.5 billion hectares produce? We know from Mike Mason's research that CAM plants can capture 1 per cent or so of the sun's energy and convert it to carbon-rich molecules in the leaves and stalks. After digestion to create a biogas – in which some energy value is lost – and then the burning of the gas in a turbine that is only 40 per cent or so efficient, we get to a figure of about 0.20 watts of continuous power from each square metre of land. That doesn't sound much and it is a lot less than solar PV spread on the same area. To power even a very efficient small LED light bulb you need a watt or more. However, if you multiply 0.2 watts by 2.5 billion hectares you get to an average power output of about 5 terrawatts.

Five terawatts is almost a third of the total energy consumption (not just electricity) of the world today. Even in a a few decades' time when we are providing enough electricity and other power to meet people's reasonable needs around the world, it will account for around one sixth of global demand. In other words, even a very inefficient process (photosynthesis) can still make a real difference to world energy supply.

Of course, the world will not easily switch all its semi-arid land into growing CAM plants and then build many millions of tiny digesters next to the new fields of Prickly Pear or other crop. But the obstacles are less overwhelming than you might think. Mason's plan is to show that digestion of CAM plants will produce good yields of energy across otherwise unproductive land and then drive down the cost of small gas turbines and the steel structures used for anaerobic digestion. Getting 1 terawatt of power out to balance the daily swings in PV output in very sunny countries looks hard work but achievable. That's more than the total electricity output from gas-powered conventional power stations around the globe, and almost as much as coal generation. It would require about 3 or 4 per cent of world land area, if I have done my numbers correctly.

We cannot know all the consequences of converting large areas of semi-arid land to cultivating CAM plants. Humankind has already much experience of the unintended, and sometimes highly adverse, consequences of converting land from one type of vegetation to another. However, CAM plants do not generally need fertiliser and are usually benign inhabitants of otherwise unproductive land. Improving the vegetation cover across substantial tracts may actually moderate the local climate, making it slightly less susceptible to extreme temperatures and flash floods. It might even be possible to restart conventional agriculture in some areas because of the soil improvement arising from using AD digestate as a soil conditioner for several years.

Those CAM plants in Tropical Power's fields are stored energy. In countries such as Kenya that will get most of their power from PV, biomass can be a perfect complement to the daily variability of solar electricity.

Dry biomass: Entrade

In June 2015 I went up to Cheshire, in the north of England, to see one of the world's first commercial small scale biomass gasifiers. Sitting inside a standard shipping container, the Entrade E3 plant from Germany takes in small pellets of woody biomass and heats them in the absence of air to several hundred degrees. At the right temperature, these pellets turn into hydrogen and carbon monoxide ('syngas') and this combination can be burnt in a gas turbine to generate electricity.

Entrade is aiming this product squarely at village and other small communities in the developing world. The owners of an E3 gasifier can utilise pellets they have made themselves. All forms of dry biomass, such as straw from wheat or twigs from harvested wood, can be used to make the small cylindrical pellets for the gasifier. Very dry material – with less than 10 per cent moisture – is ground up into a dust, compressed and then heated. At a high enough temperature the lignin in the woody material begins to melt rather than burn. When it cools back into a solid the lignin ties the whole pellet together. This is what goes into the gasifier as the fuel to make the 'syngas'.

People have been trying for decades to makes small scale gasification work and the E3 is possibly the first plausible contender, with well over 20 per cent of the incoming energy value of the pellets being converted into electricity. As well as producing electricity, it generates heat. Much more heat, in fact, than electricity. In many parts of the world, it will be possible to use the E3 to complement solar PV. As the sun fades, the gasification plant can start producing electricity by gasifying biomass from local wastes, such as corn stalks or small tree branches, and then burning the gases.

Previous small gasifiers have been expensive and unreliable. The Entrade people aimed to build a plant that is simple, cheap and works at a very much smaller scale than existing centralised power stations and can operate reliably in countries without sophisticated engineering support. It might take 30,000 of these gasifiers to replace a conventional large gas-fired power station but they can be used to make small communities entirely self-sufficient in the production of heat and electricity alongside PV.

Entrade works best with very dry agricultural wastes, perhaps including straw from the harvest in low rainfall areas. The E3 will probably be used extensively in tropical areas where there are great quantities of otherwise completely unused biomass, including stalks and straw. Many of these areas are likely to have a need for cooling rather than heating and a process called absorption chilling can be used to take surplus heat to drive refrigeration equipment. In places in the temperate world, the heat coming from the plant can be used to keep greenhouses up to the correct temperature, or even for district heating systems for homes.

Chapter 5
Demand and supply

When solar power becomes a large part of total energy supply, we'll face two related problems. The first is that PV output will only occur during daytime hours. The second is that the varying PV output will cause periods when it is difficult to align electricity supply with demand.

The next chapter will look at how we will use energy storage in batteries and other media to help deal with these issues. Storage is a way of shifting supply across time periods so that it more closely matches energy demand. This chapter is very different; it focuses on the value of managing demand so it fits more closely with available energy. This may be the simplest and cheapest way of responding to temporary mismatches between supply and demand.

The unusualness of energy supply

The provision of energy has always had one defining characteristic: we expect to have as much as we want, whenever we want it. Unlike a plane ticket, for example, whose cost might vary sharply between busy periods and the times when few want to travel, energy is usually sold at the same price across the year. It's always available and almost always sold under the same terms. Demand is never choked off by an increase in price.

Creating an energy system that always adjusts supply to match demand has been a costly endeavour. Every aspect of our grids has been designed to meet the maximum demand that will ever occur. And because prices are generally kept constant over the day and through the year, there has never been any encouragement to consumers to use less electricity or other energy source when supplies are tight. To imagine how expensive this has been, think about how many aeroplanes and crew would be needed if governments decreed that a seat had to be available to go anywhere you wanted at a standard price every hour of the year.

Electricity policy in the UK, for example, is engineered to ensure that there will be enough power at around 5.30pm on the coldest weekday in December – the time when power need is at its maximum. By contrast, hotter countries demand peaks in late summer afternoons when the air conditioners are on. Billions of pounds' worth of generating assets are waiting for these peak times just in case they are needed and many of these plants will actually generate electricity for less than a hundred hours a year. In the developed world, as energy demand drifts slowly downwards, those expensive assets are likely to be used less and less each year.

Policy makers are beginning to realise the folly of overbuilding large scale power plants just to meet peak demand for a few hours a year. Their first response has been to investigate how much small and private electricity generating capacity exists that could be used instead. A hospital, for example, has to have diesel generating equipment on site, ready to start work almost instantly if mains power fails. It may make more sense to pay to start up the hospital generating sets on the few hours a year when electricity is in really short supply. Third party companies have put together large portfolios of numbers of different private generators, and have sold this potential capacity to operators of national grids. To the surprise of many, the amounts of electricity that can be generated using small generators dotted across the branches of the main power networks are large. In some countries it may be 10 per cent of peak demand. That's a large number of new power stations that don't now have to be built only to sit idle for 95 per cent of the year or more.

Getting hospitals, large army bases or hotels to agree to switch on their diesel generators to reduce their demands for external power at peak times is clearly progress. One of the main market participants, Kiwi Power, has 650 sites in the UK that can switch on local electricity generation when the grid is under stress. However, most of those places will be using fossil fuels to create that electricity. A diesel generator emits a lot of CO_2 – often much more than a large conventional power station – so it isn't a great solution.

Much more is required. In the jargon of the industry, we need to implement large scale 'demand response' that cuts electricity use when it looks as though supplies are getting tight. Not only will this mean that countries such as the UK don't have to invest in barely used assets but, importantly from our point of view, variability in electricity supply will be much easier to manage. When the wind isn't blowing as hard as expected or the clouds come in earlier than forecast, we need 'demand response' to kick in to automatically reduce the amounts of power used by electricity users, large and small.

That's where REstore sees its market.

REstore: helping systems accommodate more PV

Pieter-Jan Mermans flashes with irritation when he talks about some of the press coverage of his business. REstore is a Belgian company running a highly automated computer platform that cuts the electricity supply to industrial and commercial companies when power is in short supply. 'They accuse us of making European industry uncompetitive,' he says, 'because they think we pay businesses to stop operating. That's not what we do at all – companies voluntarily agree with us to cut their electricity consumption for short periods. And they do this because it makes them more profitable not less. The best and most efficient companies employ us to make them money.'

He goes on to explain what REstore actually does. The operators that control the main electricity grids around the world have to keep the supply and demand for electricity closely aligned every second of every day. If they fail, even for a few seconds, the 50 Hertz frequency of the network might swing outside safe limits. This can mean that delicate electronic machinery either fails or ceases to work properly. Balancing the electricity grid means continuously watching how the level of demand from homes, offices and businesses is changing. At moments when it looks as though electricity will be in short supply the operators require idle generating stations to start up and begin to produce power.

These electricity generating plants use large amounts of fuel to ramp up power quickly. The power stations pulled in to supply the extra power at very short notice often only work a few tens of hours a year. So it is expensive and inefficient – and this rapid ramping up and down of production can create six times as much CO_2 per unit of electricity as a conventional power station. Mermans also noticed that flexing electricity supply was likely to get more and more difficult as the electricity industry evolves and the amount of unpredictable solar and wind on the grid continues to rise. Equally importantly, simple gas turbines – essentially jet engines – which are the fossil fuel power stations with the greatest ability to respond to urgent requests to produce more or less power, are being phased out. He realised that in future it might be easier to focus on adjusting demand, not supply, in order to balance the electricity market.

It's fair to say that people who'd been in the electricity industry for decades would have been very sceptical about his business judgment when he decided on his strategy; conventional wisdom was that industry and commerce cannot easily reduce their energy consumption and that flexing the supply of power is always much easier. Nevertheless Mermans and his business partner Jan-Willem Rombouts created a company that allows large power users to promise to reduce their electricity use for minutes or hours in return for cash payments. Unlike players such as Kiwi Power, which harnesses highly polluting diesel generators to build up the supply of electricity, REstore's business lies in genuine demand reduction, providing an important – but strikingly little understood – innovation in energy markets.

Mermans comes from the energy business and knows how the electricity markets operate while Rombouts arrived from Goldman Sachs where he'd specialised in real-time management of large portfolios of financial assets. Over a period of about two years, they built one of the first software platforms in the world that allows participating companies to offer to genuinely reduce their electricity demand when required. The financial returns can be substantial. For example, if you can guarantee to cut your energy use by a predetermined amount within a few seconds, National Grid in the UK might pay you tens or even hundreds of thousands of pounds a year.

Of vital importance, each organisation sets the rules for when and if it will be prepared to cut its power use. With hundreds of different participants, REstore can be highly confident that sufficient numbers of its customers will be happy to reduce their electricity use when required.

They now operate an entirely automated market in France, Belgium and the UK in which hundreds of organisations make commitments to reduce their electricity demand at very short notice in return for a yearly payment. Their platform has many similarities to some of the marketplaces that Rombouts had been working on at Goldman Sachs.

You might imagine that cutting electricity use by worthwhile amounts at very short notice is extremely disruptive to the typical business. Mermans convinced me otherwise. The first example he gave me was from paper mills, which use huge amounts of energy in their day-to-day operations. A large paper factory first creates a semi-liquid pulp and stores in a tank. It then makes paper by passing the pulp through a long run of expensive machines. Both processes require a lot of electricity. But for those operating the plant the most important thing is keeping the second phase, paper-making equipment, fully utilised, twenty-four hours a day. So they always ensure that there's a good store of pulp, waiting in a large pool ready to be fed into the machines that press it into paper.

That is where the opportunity lies for the paper company to gain from participating in REstore's marketplace. The stock of pulp is a buffer. If no pulp is made for a couple of hours, there'll usually still be enough to feed into the paper machines. As long as the stock is at normal levels, the paper factory will be happy to lose the electricity supply to their pulp-making machines. The REstore system is given access to the sensors in the pool of pulp. If the levels are high enough, it will have permission to turn off the electricity temporarily. If they are not, its system temporarily removes the paper factory from the portfolio of demand response participants. 'Every business has bottlenecks,' says Mermans. 'Every production line has points at which goods are stored, waiting to go through that bottleneck.' The machines that make those goods can usually be shut down for a half hour at no cost to the business whatsoever. 'In fact, we help them reduce their electricity bill and they get paid for cutting usage when power is in short supply.'

A couple of months later, another UK demand response coordinator, Open Energi, made a similar point by publicising some of the details of its contract with Tarmac, a leading asphalt producer, to introduce automated management of the electricity demand of its 200 asphalt tanks. Asphalt is made by melting bitumen and mixing it with gravels to form a liquid that can be used to restore roads or create waterproof surfaces. The tank in which the mixture is made needs to be kept very hot using electric heating but the amount of power used can be varied. As Open Energi said:

> [Our] technology automatically and invisibly adjusts power consumption of these bitumen tanks to help manage fluctuations in electricity supply and demand. [Demand response] is ideal for use with stored energy devices such as bitumen tanks, because although they need energy, as long as they operate between expected temperature limits, it does not matter precisely when that energy is used.

It's also worth noting that Tarmac saves money by doing this, as well as reducing its responsibility for carbon emissions, just like the companies whose processes are managed by REstore.

Pieter-Jan Mermans and I discussed other examples of when electricity use can be avoided for short periods because the user has a protective buffer. A domestic fridge isn't actively working all the time and it can be turned off for half an hour with no effect on food condition. Or, at a much larger scale, a frozen food warehouse could reduce the electricity demand for its chillers, probably for several hours. A hotel might choose to limit its air conditioning demand for shorter periods.

One of the main targets for demand response companies is the waste water treatment industry. Sewage farms and their associated infrastructure use large amounts of power. They can usually stop some of their machinery working for long periods without any adverse effects on the overall operation of the plant. Enbala, a demand response company based in Vancouver, Canada, that operates in a similar software-driven way to REstore, says it can work with 'anything that has a motor' to reduce electricity demand for an hour or more.

Steve Connolly runs Arriba, a Cambridge, England, start-up which focuses on manufacturing more efficient heat pumps. He told me that about 30 per cent of all industrial demand for electricity is used for air conditioning and refrigeration. The use of power for this equipment could be modulated to match short-term changes in the availability of electricity. His aim is to help his customers precisely match the minute-by-minute supply of locally generated electric power, such as from PV arrays on their roof, to the electricity demands their machines are making. Like an increasing number of others, he thinks that many buildings will run internal low voltage direct current grids, taking the DC from solar panels or wind turbines and using it directly in the business without ever converting it to alternating current. This will both save money and also reduce the problems incorporating solar PV and other renewables into the electricity supply.

However, no industrial company can absolutely guarantee that it will want to reduce its demands on the electricity grid at a few moments notice. That's where the cleverness of the REstore approach comes in useful. By having hundreds of different factories in its portfolio, and highly sophisticated sensors working at each stage of manufacturing processes in each plant, REstore knows at every second of the day which of its customers, and indeed which machines within each of these businesses, can cope with having their electricity supply cut, and for how long.

REstore is reliant on the law of averages. Clever use of statistics allows it to estimate the minimum amount of electricity use it can cut from the large range of different plants in its basket. One of its biggest customers is the world's largest steel company, Arcelor Mittal, a business with a total worldwide energy need approximately equivalent to that of Austria, or about 25 per cent of the UK. An electric arc furnace that melts recycled steel uses a prodigious amount of electricity to heat the metal. Much of the time Arcelor Mittal wouldn't want to lose its power supply. But once the steel is melted and the impurities start floating to the surface in the furnace there's enough of a buffer of red-hot heat for the electricity to be shut off for a few minutes. Arcelor Mittal says it saves money, and reduces its carbon emissions, by allowing REstore to shut off the power occasionally.

In total, REstore now has over 1 gigawatt of power consumption signed up around northern Europe to participate in near real-time demand response (that would meet about 2 per cent of peak UK electricity need). I asked Pieter-Jan Mermans what might be the maximum amount of short-term demand flexibility from industrial users. He responded that his business 'continued to be surprised' by how much capacity is being offered to him. Looking at France, he continued, demand response from industry looks as though it might be at least 6 per cent of peak demand. In the regions of the US controlled by the grid operator PJM, it was as high as 12 per cent but this included companies who switched to diesel generators as well as those who actually reduced their load. I think we'll probably find that the eventual figure is much higher than these estimates.

Cutting power demand in the home

When the 'Internet of Things' comes fully into operation, every single machine in the world, ranging from domestic radiators to steel furnaces, will have the capacity to be turned off at short notice. I'm not saying for one minute that people will voluntarily agree for their usage to be curbed but it will be theoretically possible to delay washing machines, ask laptop computers to run on their batteries, or turn off dishwashers. In return, the household will see reductions in bills, or at the very least bills won't be as high as otherwise.

Once the electric car has become commonplace it will provide a large increment to the flexibility of the electricity system. Today's Nissan LEAF charges at a rate of about 3 kilowatts in a domestic garage. Much of the time the car will not be charging, either because it is away from home or because its batteries are full. But the 10 or 20 per cent of cars that are charging at any one time can be temporarily disconnected to help reduce demand. If the car owner agrees, vehicles with full batteries can discharge to the grid, reversing the usual flow. (Most electric cars cannot yet send electricity back into the electricity system, but this capability will be standard on new electric cars fairly soon.)

You might imagine all this is years away. But in California the start-up OhmConnect provides an app for smartphones that tells customers when the electricity market is short of supply and wholesale prices are beginning to spike upwards. (This can be at the daily peak of demand but is more likely to arise when supply is unexpectedly curtailed, such as when a power station fails.) The company sends out an alert, saying that the next hour will be an 'OhmHour', and it will pay its subscribers to reduce their electricity draw for 60 or 120 minutes.

How does it earn the revenue to pay for this? It sells its promise to reduce demand to the Californian wholesale electricity market. A megawatt hour saved by OhmConnect is identical to a megawatt hour generated by a highly polluting power station that is turned on especially to capture the very high prices paid at times of stress.

When they get the alert, customers can then choose to reduce their demand by turning off air conditioning or pool pumps, which are big users of electricity, or any other appliances in their home. Ninety per cent of Californian homes have so-called 'smart' meters that collect information on electricity usage minute by minute and send it back to utilities or companies like OhmConnect. The app looks at the pattern of usage over the last days and weeks and estimates whether the homeowner has run the house with a lower than expected amount of electricity during the 'OhmHour'.

If so, a payment is made; the user gets 80 per cent of the price that his 'negawatts' (negative watts) have been sold for and OhmConnect keeps the rest. The company says some of its users have made more than $200 in the past year from cutting their electricity use every time the alerts have been sent. More typical cash rewards have been about $100. This is not a big sum and not worth the time and inconvenience to get. However, if you live in an area that is subject to a substantial risk of blackouts, as parts of California are, then you might see a real social value in joining with thousands of others to reduce demand at critical times. According to the company, many of its customers also use their payments to club together to support local charities and schools. Participation in the OhmConnect system seems to be more of a communal activity than a conventional money-making venture.

The next phase in OhmConnect's development is a partnership with the French electrical equipment manufacturer Schneider Electric. Schneider makes controls for domestic electrical appliances, such as thermostats and wall plugs, that can turn current off when given instructions via WiFi. An OhmConnect customer can install controls for all the electrical equipment in the house, meaning that when a price spike arrives electricity use is automatically cut back to nearly nothing, and this time without any intervention from the homeowner. As you'd expect in California, the home of Tesla, there's special equipment that can stop electric cars charging in periods of stress on the grid. And because a car plugged into a socket on the wall might be using 90 per cent of a house's electricity consumption this will really make a difference. The average returns will be greater.

Apps like OhmConnect will allow real-time matching of supply and demand, as well as helping to smooth out peak demand. REstore provides demand management for large industrial users and OhmConnect shows how it can be done in homes and small offices as well.

Demand response: what happens next?

REstore provides a good example of how modern information technology can be used to shift demand away from time periods of stress in an electricity system. The incentive used is a payment to those member organisations that stop using power. The obvious other step is to charge customers more when demand is high or supply is limited. Around the world, 'time-of-use pricing' is becoming more common, although at a lamentably slow pace. During periods when electricity is in short supply, the price is higher (a feature of almost every other market in the world, after all).

When solar power is the most important source of electricity, it will be logical to charge far more at night than in the day. This will probably have two effects. First, power demand will be shifted from the night to the day. And second, if incomplete case histories around the world are any guide, there will be a small but measurable reduction in total electricity use as people simply decide they can go without using some of the electricity that they had used in the past.

The Canadian province of Ontario is a pioneer in time-of-use pricing. In late 2015, the local utility charged twice as much in peak time as it did off-peak. The time of the highest prices varies during the year, with the summer peak during the day and the winter maximum in the morning and early evening. Of course, this reflects the pattern of demand, with air conditioning creating high demand in summer afternoons and heating and lighting dominating in the winter. Ontario is also largely reliant on hydro-electric power which cannot always ramp up and down swiftly. Careful Canadian studies have shown small but measurable switches in electricity use away from the afternoon peak in summer. Peak demand is between 3 and 4 per cent lower than it would otherwise be if prices were the same all the day. Interestingly, in winter electricity use declined over the whole day, not just the peak and there was little, if any, switching between periods of high and low prices.

How does a utility such as the one in Ontario know when homes are using a lot of electricity? The answer is that it will have to install smart meters for all homes. Similar devices are gradually being put in UK homes to prepare for time-of-use tariffs. These meters can be set up to measure the amount of electricity used at differing times of the day and therefore calculate bills using the tariff for each slot. A householder or a business might be offered very low prices when the sun is shining or the wind is blowing

Some of the same effects on electricity consumption that were observed in the Canadian studies were seen in a survey of 1,000 domestic electricity customers in London in 2013.

The organisation running the trial offered three different rates, ranging from 4 pence per kilowatt hour at cheap times to a maximum of 67 pence at times of serious problems lasting several hours with the electricity supply. (These were all mock emergencies, invented for the purpose of the experiment.) If behaviour didn't change, the total bill would be the same. On average households reduced their use of electricity by about seven per cent at times when the 67 pence rate was being applied. The most responsive quarter of households managed around 20 per cent. As might be expected, the periods of very low prices induced some increase in demand, although the percentages were small. In the future of a renewables-dominated grid, it may be almost as important to be able to increase demand at times of electricity surplus, although the vital innovations dealt with in the last chapters of this book may enable network operators to productively use that excess to create stored hydrocarbon gases and liquids.

The households participating in the London trial had to make adjustments, such as switching off the electricity-guzzling tumble drier, themselves. In the near future, this will all be automated – if the customer agrees, of course. In this study, even without this automation but just simple notification by text messages, about 95 per cent of the triallists achieved lower bills as a result of accepting time-of-use pricing. The report of how the trial worked also says, perhaps surprisingly, that most similar experiments around the world have shown that around 80 per cent of participants enjoyed 'playing' in the trial.

In early 2016 the Cornish town of Wadebridge launched the first commercial trial of variable pricing in the UK. Households were offered a much lower price for the daytime hours of summer when PV is likely to make wholesale electricity cheap. They will pay 5 pence per kilowatt hour, less than half the usual rate, between 10am and 4pm. But outside these times the price will be 18 pence, over three times as much and well over the standard price for electricity. Unless the house has a battery to store electricity, or the residents are astonishingly clever at ensuring all their electricity use is compressed into a six hour period, this will not save money. But it is a precursor of how electricity pricing will work in a decade's time as electricity becomes similar to other commodities for which pricing is used to match supply and demand.

With full automation – involving all the appliances in a house being linked up through the internet to a control centre – the short-term responsiveness of electricity demand to price signals will be greater than today. There may be some resistance to this change, but within ten years some of us will load our dishwasher and let the grid operator choose when to start it. If necessary, and when we are prepared to pay a premium, we will be able to override the decisions made for us. We can expect that all non-time-sensitive electricity use will be eventually pushed into periods of peak renewables output. The pleasant consequence will be that electricity bills go down, not up.

Faced with unusable quantities of PV generation during the day, electricity companies around the world will probably eventually want to introduce a peak charge in the evening that is three or even four times the price at the times when supply is abundant during the day.

Hawaii – with more PV capacity per head than almost anywhere else, no large regional grid to rely on and expensive electricity produced by diesel generators – provides us with a postcard from the future. Oahu, the biggest island, with more than 10 per cent household PV penetration, has a proposed daytime tariff of 13.4 cents (about 9 pence) per kilowatt hour from 9am to 4pm when solar is abundant. From 4pm to midnight the rate rises to almost three times as much – 38.5 cents (about 25 pence), when sunshine reduces to zero and demand for air conditioning is highest. The intent is clear; customers are being pushed into using as much power as possible when solar generation is at its peak.

The secondary effect of this tariff gradient is to greatly encourage the use of domestic batteries to store power from the PV array during the day and then to employ that electricity at the time of peak prices. Home storage systems don't make much financial sense without time-of-use prices. But with Hawaiian electricity rates, the logic for installing a domestic battery, such as the new Tesla Powerwall, becomes almost overwhelming for those who can afford it.

Countries with year-round sun and relatively little energy-guzzling air conditioning will see electricity prices that are low during the day and peak just after the sun goes down. In the UK, prices will be low in the summer and highest in early evenings during the winter. In California, the relatively short days of spring and autumn already see solar-generated electricity falling away before air conditioning demand begins to tail off. As PV grows in importance to the electricity grid, the sharp rise in unmet demand around 5pm will need to be moderated by time-of-use pricing and, as plans currently stand, this will be introduced in the state by 2019.

The chart below shows the 'California duck' curve. The demand for electricity generation from fossil fuel sources dips sharply as the sun rises in the sky. As the afternoon wears on and PV falls away, the need for other sources of power rises very sharply indeed. The 2020 forecast – which assumes far more solar electricity generation than in 2015 – suggests a doubling of electricity demand in the space of four hours from 4pm to 8pm on a typical spring day. This precipitate rise will have to be moderated by time-of-use pricing.

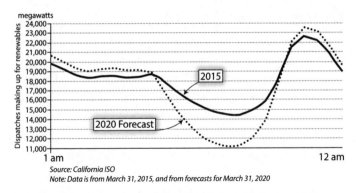

Source: California ISO
Note: Data is from March 31, 2015, and from forecasts for March 31, 2020

As more and more solar power is installed on the California grid, the demand for other sources of power falls sharply in the middle of the day.

Energy efficiency and demand response

Some of the resistance to renewables comes from a deep-rooted feeling that instant access to all the power we want is close to a human right and any attempt to mould usage to availability of power is an infringement of liberty. This appears to be the case even if management of demand is through the mechanism of varying prices, and not administrative diktat. Strangely, many of the people who resist the move to variable energy pricing are of the political right and have an otherwise fierce belief in the value of free markets.

It is in the nature of such markets to use changes in price to bring supply and demand into equilibrium. For example, the price of a Eurostar ticket from London to Paris is low at 6am on a Sunday in order to encourage price-sensitive people to travel at that time. Indeed, train fares can vary almost three-fold within a day. Few complain about this. Similarly a robust electricity system needs flexible pricing that curtails or expands demand as the supply of power changes. More and more electricity use will have to be concentrated in the daylight hours in a world dominated by PV.

The importance of lighting

Brenda Boardman and I are both trustees of a charity that uses its funds to reduce the number of households in 'fuel poverty' in the UK. Fuel poverty afflicts households that find it difficult to pay for enough heat and electricity to be comfortable. She is also one of the UK's most influential researchers on domestic energy efficiency and I wanted to talk to her particularly about reducing electricity use in lighting. My logic was that lighting tends only to be used when solar PV electricity isn't available so if we can cut lighting needs we can make a real difference to the need to store electricity.

The UK is a fairly typical high latitude country in that the daily peak electricity demand tends to occur at around 5–6 pm in the winter. People are arriving home and turning on lights and appliances and cooking dinner before offices and factories have closed for the evening. This means that we see a hump in electricity demand until businesses have finally shut.

The sharp evening peak is an important problem to solve. Solar PV cannot provide the power without an expansion of electricity storage. The sharp 20 per cent increase in UK electricity demand between 3pm and 5pm isn't always easy to

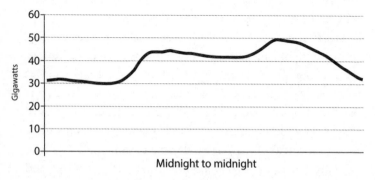

How demand for electricity varied in the UK on a typical day in January 2016.

handle, even today. The short peak requires large numbers of expensive power stations to stand ready to operate just for a few hours each winter day. If we can level out electricity usage by reducing the need for lighting we decrease the requirement for fossil fuel generators to operate during these daily peaks. In doing this, we also make it less financially attractive to build new fossil fuel power stations that typically make all their profits in these peak hours.

Brenda recently published a paper on how much the UK could reduce electricity demand during the winter early evenings by improving the energy efficiency of lighting. She believes the numbers are large. On a cold December afternoon at the point of maximum demand the UK might be using 52 gigawatts of electricity. In her work she showed that lighting, both domestic and commercial, was as much as 16.75 gigawatts of this peak-time total, or almost a third.

She thinks that by switching to ultra-efficient LED lights this can be reduced to no more than about 2.5 gigawatts by no later than 2030. This six-fold reduction would reduce the UK's peak demand by at least 10 gigawatts and would largely remove the hump in early evening usage. This has obvious importance for the rise of solar because almost all the reduction in lighting use will occur, for obvious reasons, at times when PV power isn't available. Taken across the year – not just in the peak times in winter – electricity demand from lighting is almost a fifth of the UK total, Brenda writes. She thinks this will fall to little more than 3 or 4 per cent, reducing total electricity use by over 15 per cent.

Is it reasonable to expect that we will see this huge cut in usage? The best LEDs are now more than twice as good at turning electricity into light as compact fluorescent lamps and several times better again than the halogen bulbs now used in kitchens, bathrooms and living areas in many homes around the world. The switch to LED is happening fast as prices tumble and major retailers such as IKEA stop selling any other type. Of course, things never quite turn out as you expect – people might stop turning lights off when they leave the room because it is so cheap to leave them on – but the rapid advance of LEDs will almost certainly cut electricity demand substantially at periods when the sun isn't shining.

Jason Palmer at Cambridge Architectural Research and his colleagues have done very detailed work on the makeup of household electricity use. In a 2013 report they showed that fridges and freezers were responsible for about 10 per cent of household energy use. This is less than the 30 per cent of all electricity used for refrigeration in commercial buildings, but still a large amount. Add in 'wet' appliances such as washing machines and dishwashers and water heating by electricity and the total increases to about a fifth of all electricity employed in homes. Fridges and freezers can lose power for an hour with no impact on safety or the quality of the products stored. Washing machines and dishwashers can be stopped and restarted with little or no impact. All of these machines can be turned off remotely for an hour to deal with a temporary shortage of electricity. And there is no reason why users should not be offered the right to override the central instruction if necessary. Demand management programmes that incentivise householders (and businesses) to turn off refrigeration plant during the two or three hours of early winter evenings will further assist the flattening of demand.

Heat pumps: Energiesprong

Most of this book has been about generating a secure supply of electricity without burning fossil fuels. When we have done this, we will also be able to convert almost all surface transport to electricity and in colder countries we will switch heating away from gas boilers to electric heat pumps, a form of reverse refrigerator that heats the internal air in houses.

So far, heat pumps have been a near-disaster in the UK for most of the people who have installed them. Heating bills have usually risen and the pumps have been poor at making homes warm. An important part of the reason has been the woeful insulation of most British housing. Any attempt to build a renewables-based future in places like the UK and some other places in northern Europe will fail unless combined with massive reductions in the heat losses of almost all homes, and many commercial buildings.

The average home in the UK uses over four times as much energy in the central heating boiler as all electric appliances and lights combined. Decarbonisation of heat is a harder challenge than creating a renewables-based electricity supply. Even if we invested in a huge oversupply of solar panels, countries in northern Europe could not easily supply enough heat from electricity to keep people warm in midwinter.

What do we do? As well as enforcing the highest standards on new construction, we urgently need to improve older buildings. Partial programmes, such as better insulation of the walls of older homes, have had little effect on the overall energy use of the buildings. One UK government survey suggested an average improvement of no more than 15 per cent in energy use when a typical pre-1920 British house had insulating cladding installed on its external walls. Other figures are equally unimpressive; adding more insulation to the loft cuts energy requirements typically by only about 2 per cent. UK government policies to improve insulation have been expensive and largely unsuccessful.

Dutch company Energiesprong thinks it knows how to improve on the indifferent record on insulation in many colder countries. Energiesprong insists on whole-house renovations that move the house to having energy consumption less than zero over the course of the entire year. The PV on the roof provides more electricity than is used by the house. The high quality insulation of all the walls and roof removes the need for a conventional central heating system.

Energiesprong doesn't do the work itself; it acts as a marketplace between property owners (usually institutions that own social housing) and construction companies that it has trained to do full-scale refurbishment to carbon neutral standards. The Energiesprong vision is simple: every aspect of a house's energy performance needs to be improved. This means major structural changes, including replacement of the roof. As you'd expect, both sides of the new roof are built out of solar panels on a steel frame. Even on a north facing roof this makes sense because panels are no longer more expensive than traditional building materials.

As importantly, Energiesprong wants its efforts to result in a much more desirable house than before renovation. The refurbished home should look better and more modern as well as being free of damp. It thinks that once the obvious benefits of Energiesprong home improvements are seen in a street, it will be easy to persuade other families to demand better housing. This is a new approach and my sense is that this is absolutely what is needed to create a movement for energy efficient housing. Saving money, or reducing carbon emissions, isn't enough; the house must also look and feel better if we are to encourage people to back major refurbishments.

Crucially, all Energiesprong components are built off-site and can usually be installed within a week. Project director Arno Schmickler told me that the disruption to homeowners is minimal because almost all the work can be done without the construction workers going into the house.

A Dutch Energiesprong house, before and after refurbishment.

Energiesprong's work has now resulted in over 700 successful home improvements in the Netherlands. Typically, energy efficiency is improved by so much that the typical renovated house no longer needs a gas boiler or other heat source. It switches all the lighting in the home to LEDs, of course, and improves other obvious aspects of the house's energy performance. It usually works with social housing providers because these organisations often have large numbers of houses of similar design so that it can get some major cost reductions.

In the UK, where it has just opened, Energiesprong is also working with social housing associations and major builders. At the end of the refurbishment, the tenant gets a warm house and a guaranteed amount of free electricity and hot water. In return the occupier pays a monthly fee, intended to be less than the utility bills before the conversion. So the occupier will also usually save money.

Of course, my first questions to Arno Schmickler, who runs the UK operation of Energiesprong, were about the cost of the work needed to transform the performance of an older building. The average gas and electricity bill for a small house in the UK is about £1,000 a year. Does insulation actually make any financial sense?

Arno was quite open about the large bills for the first houses that any of its partner companies convert. Energiesprong and its construction affiliates are only just moving out of the prototype phase and the builders are learning how to do the work cheaply. Initially, whole house improvements were costing about €120,000 (just over £90,000) for typical smaller homes. One of the main Energiesprong building contractors in the Netherlands is hoping to cut this figure to no more than about €35,000 within two to three years. This is still a large sum but it is easier to justify if the social housing company that owns the home would have needed to do a full refurbishment anyway.

A few years ago construction professionals would have laughed if you'd suggested carbon neutrality could be cheap enough to make financial sense. But Energiesprong's early success means that many building companies now want to join the Dutch and UK programme to get in early to the scheme. As with solar PV, the interest rate charged on the money spent to improve a property is absolutely critical. In the Netherlands, Energiesprong partners are able to get government-backed financing for its schemes, paying around 2.2 per cent a year for the money. Spread over thirty years, this should mean that both the tenant and the housing association are better off. England is different. Housing associations face much higher rates of interest, possibly more than twice the level of their Dutch equivalents, because they cannot tap into public sector sources of finance. It will be more difficult to make the numbers work and Arno Schmickler thinks it might be that Scotland – which is more determined to improve the quality of its public housing – will be the first place in the UK to really push for large scale carbon neutral housing.

The family renting a home that Energiesprong refurbishes is, in large part, getting a new place to live: warmer, less draughty, and often with a refurbished kitchen and bathroom. The outside is glossy and well finished. To most eyes, the house is a much better-looking, more desirable place to live. Energiesprong sees this as the real incentive for residents, whether owner-occupiers or tenants. They'll get to live in a more valuable house, as well as one that costs nothing to heat or light. If they own it, it'll be worth more and be easier to sell. Arno told me that Dutch mortgage banks are happy to lend more on houses that have been through an Energiesprong refurbishment.

It's obvious that in the long run these houses will also benefit from having batteries to store surplus electricity. The Energiesprong estimate is that many houses can justify a 10 kilowatt hour battery. With that installed, the typical house will move from 20–25 per cent electrical autonomy to 60 per cent or more over the course of a northern European year. (And there will be no need for central heating at all.) That means that most electricity generated on the roof will stay in-home, dramatically reducing power taken from the grid.

It may not be as glamorous as the other innovations in this chapter but in cold countries we need to add dramatically better insulation of homes as the third leg of the energy transformation to complement the use of renewable electricity and large scale storage. Without this, the solar revolution will only ever be a partial solution.

Chapter 6
Battery power

Electricity users can adjust their demand in response to variations in the availability of power – as we have seen in the last chapter. But even in the sunniest countries, this won't be enough. If PV is to take over the dominant position in the world's energy system we need large amounts of storage of electricity, both for short and long periods. The good news is that batteries increasingly look as though they will provide most of the electricity needed for overnight use, at a cost far less than seemed possible just a couple of years ago.

An autumn 2015 report from credit rating agency Moody's noted that battery prices have already halved since 2010. All the evidence from around the world is that they will continue to fall for the next few years at rates that might even match PV. Batteries are where PV was five years ago: a technology getting more competitive every year that is still not quite cheap enough for mainstream applications. But change is happening fast. By the end of 2016, another 5 or 10 per cent will have been hacked out of battery costs and another set of applications will be starting to use electricity storage.

Although the lithium ion battery cells used in electric cars are attracting the world's attention, it is probably a mistake to assume this is the only type of electricity storage that will succeed. Lithium ion may be best at addressing needs for batteries that can discharge very rapidly, giving a surge of power when needed. They can deliver, when required, a large number of kilowatts compared to the total number of kilowatt hours in store: like a bath of water that cannot hold very much but which has a large plug hole. Cars need this sort of power.

Other types of battery will be required to provide a consistent power level for many hours, perhaps during the night hours in order to power an off-grid village. It's my guess that a completely different type called a 'flow' battery will meet this need. Flow batteries can hold a lot of energy but only dispense a relatively small amount each hour. By contrast to the lithium ion batteries, this is a large bath with a smaller outlet.

Lithium ion

We can thank the car manufacturers for much of the recent progress in lithium ion electricity storage. These companies and their suppliers are betting on the brute force of the experience curve. Build bigger and bigger factories and move into new markets, such as domestic electricity storage, and the increases in the number of batteries produced will inevitably push down costs. It doesn't really matter whether lithium ion is the perfect way of making a battery, the important thing is that you can keep reducing the manufacturing cost by ramping up volumes produced. As with silicon, there are always people ready to say that we should be using different battery chemistries using other molecules but their views are getting pushed aside by the relentless improvement in the pricing of standard lithium battery cells.

Invented about thirty years ago, batteries using the electrical characteristics of lithium grew to significance because of their vital role in powering mobile consumer electronics, such as phones and laptops. But the recent advances in cost and performance have mostly come from the use of lithium ion battery packs in electric cars. Your mobile phone has a battery that holds about 7 watt hours (0.07 kilowatt hours) but a modern electric car may have about 30 kilowatt hours of storage. It is exactly the same type of battery in both appliances. However, the car battery has about 4,000 times as much electricity storage. Electric transportation has pushed batteries down the experience curve further and faster than laptops and mobile phones could have achieved on their own.

The single most important figure in the battery industry today is Elon Musk, the head of Tesla. Working with battery manufacturer Panasonic, Tesla is building a vast 'Gigafactory' in Nevada that will produce battery packs for Tesla cars and

Elon Musk unveils Tesla's 'Powerwall' battery storage at an event in California, April 2015.

also for the company's new residential and grid storage systems. By building a factory of this size Tesla is expected to reduce the cost per watt of storage to about half the level of just a couple of years ago. This will accelerate electric car sales and the use of lithium ion batteries in static storage as well.

The experience curve, again

As with solar PV, virtually no one predicted the declines in battery cost that we are now seeing. In 2009 Deutsche Bank reported the cost of lithium ion batteries as $650 per kilowatt hour and predicted this would halve to $325 by 2020. Slightly less optimistically, the Boston Consulting Group (BCG) said, the same year, that the price of packs would be $360–$420 per kilowatt hour by the end of this decade, and concluded that 'the cost target of $250 per kWh is unlikely to be achieved at either the cell level or the battery pack level by 2020'. (A battery is composed of many cells linked by wiring and an electronic safety and control system; a battery pack holding is always more expensive to make than the cells themselves.)

That pessimism wasn't correct. An academic paper published in early 2015 gathered as many estimates or battery costs for cars between 2007 and 2014 as the Swedish authors could find. This work suggested that the largest manufacturers of electric cars, Tesla and Nissan, were already achieving costs in 2014 of around $300 per kilowatt hour (below the levels predicted for 2020) for complete battery packs. The paper also gathered forecasts of future costs. The authors estimated that costs had fallen by 14 per cent a year from 2007 onwards and they could see no reason for the decrease to slow.

The surprises continued. In October 2015, car manufacturer GM announced in a formal presentation that its individual cells were costing $145 per kilowatt hour. Large companies like GM don't exaggerate numbers like this for fear of future litigation from shareholders. So the business is already buying cells at a price almost half that BCG thought 'unlikely to be achieved' in 2020. By 2021, GM announced, its cells are expected to be around $120 per kilowatt hour. I'd guess that this means its battery pack will be around $170.

Similarly, Tesla is looking for its cells to be about $100 per kilowatt hour by the turn of the decade. Other car manufacturers will be desperate to achieve similar levels and my bet is that the rapidly growing Chinese electric bus and car manufacturer BYD, a core holding of US investor Warren Buffett, will be producing millions of inexpensive and low-specification battery-driven autos by that date, further increasing the volumes of cells that are required and pushing the cost down to levels nobody has dreamt of, even now.

How Tesla's new Gigafactory in Nevada will look. It will deliver as many lithium ion cells in a year as the whole world produced in 2013.

A standard 2020 car will probably travel about four miles on each kilowatt hour of stored electricity, meaning that a 200-mile range will need a battery pack holding 50 kilowatt hours. That might cost as little as £5,000/$7,500 by the turn of the decade, making the purchase price of electric cars directly cost-competitive with petrol versions, in addition to being much cheaper to run, insure and maintain over the course of their life. At the prices Tesla is promising for 2020, employing lithium ion batteries for bulk storage of electricity – not just in cars – will make financial sense almost everywhere. We can already see illustrations from around the world that large scale battery storage is seen by some utilities as the cheapest way of dealing with periods of peak demand.

In April 2014, the energetic founder of Bloomberg's renewables information business, Michael Liebreich, presented his calculations on the rate of decline of lithium ion battery costs. His estimate of how much costs fell every time accumulated production doubled? He suggested an experience curve slope of 21.6 per cent, a rate of decline even faster than solar PV. Most other estimates are slightly more conservative; Michael Liebreich does tend to be a bit enthusiastic. But even if the experience curve is slightly less steep than he suggests, the rate of growth of battery volumes more than makes up for this. (Remember that the speed of decline of the cost of a battery or a solar panel is determined both by the steepness of its experience curve and the rate of growth of accumulated production.)

You may recollect that BCG commercialized the idea of the experience curve fifty years ago as part of the toolkit for its consulting practice. But the business that built a large part of its reputation on estimating future manufacturing costs got batteries spectacularly wrong in 2009 by voicing pessimism about what could be achieved before 2020. Even the BCG people, whose livelihood depends partly on forecasting the cumulative impact of production volumes, didn't really believe in the full, unavoidable, almost magical impact of the experience curve.

Quicker, simpler, cheaper: 24M

Batteries are evolving in a remarkably similar way to PV. There may be no further technical breakthroughs with lithium ion but instead a whole series of small improvements that will accumulate to major reductions in cost. Small increases in the energy density of cells will, for example, mean lower packaging costs per unit of energy capacity. The amount of electrode material per cell will fall, the speed of coating will rise and the assembly rates will increase. A multitude of tiny changes in production methods will cumulatively mean substantial continuing falls in production costs as engineers drive out all avoidable waste and make small improvements in the amount of energy delivered by each cell.

Or, just possibly, all these improvements will come at once. One Boston company intends to completely revolutionise the manufacture of lithium ion batteries. 24M, a business set up by Professor Yet-Ming Chiang of MIT, promises that its new design will rewrite the way that lithium ion cells are made. When Sony began to make lithium ion rechargeable batteries in the early 1990s to power its orginal Walkman music players, it used old manufacturing equipment that had been used to make the magnetic tape for music cassettes. That single decision determined the way lithium ion batteries have been manufactured for the last twenty-five years, even though the weaknesses of the approach were obvious even then.

Chiang's company now has a process to manufacture batteries that is quicker, simpler and requires much less capital equipment. Inside the battery, there is far less redundant

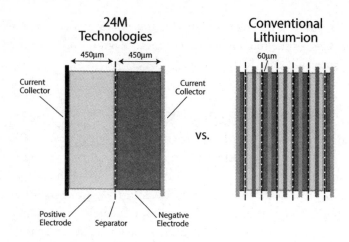

The crucial 24M advantage: 24M technologies on the left,
conventional lithium ion on the right.

material and therefore a much higher energy density. Author
Steve LeVine wrote in a recent profile of the company: 'Lith-
ium-ion cells typically contain fourteen separate material
layers; Chiang simplified them, allowing him to reduce the
layers to just five.' He also focused on the improvement in
manufacturing that comes from not having to take as much as
twenty-two hours to dry the electrodes prior to the completion of the cell. 24M intends its entire process to take an hour.

The company promises sub-$100/kWh batteries by 2020,
a target that even Tesla will struggle to beat. As importantly, it
will be possible to set up factories that are very much smaller
and less expensive. Chiang talks of manufacturing plants that
cost as little as $11 million, a minuscule fraction of the billions
Tesla will spend on its Gigafactory. For that an owner might
get a factory that makes about 80 megawatt hours of storage
a year. The market may not be electric vehicle manufacturers
but rather businesses seeking to deliver static storage to the
electricity grid for overnight use.

Domestic electricity storage

From the mainstream media, not yet interested in 24M's remarkable claims, Tesla gets all the attention, partly because of the glamour of its fabulous cars. It has entered the market for residential storage of electricity using a new product called Powerwall. This product gives a few kilowatt hours of storage so that the household can use its surplus solar power from the day to provide its electricity at night. A much larger version is also supplied for storage of far larger quantities of electricity by grid operators.

I'm far from certain that Tesla will dominate the market for domestic batteries, despite its extraordinary ability to excite media attention. Its much quieter competitor Sonnen moves stealthily ahead from its base in the south of Germany and I think its approach may be more successful. Sonnen has built its business around small domestic battery storage units holding a couple of kilowatt hours or so, enough to keep the lights on, run the fridge and watch some TV on long December nights. This amount of small scale storage makes most sense in countries with high retail prices of electricity. Why not keep your own power for later consumption rather than export it cheaply at midday and pay extortionate rates to buy in electricity in the evening? German domestic prices are as high as 28 cents (more than 20p) a kilowatt hour, almost twice the cost in the UK. The incentives to 'self-consume' your own electricity are therefore high and Sonnen has sold more than 8,000 units for homes and small businesses and has huge, well-funded ambitions to sell many more around the world.

I rang Mathias Bloch, spokesman at Sonnen, to understand in what other ways their product is better than others on the market. Their battery system, he said, does far more than store electricity from the PV array on the roof. It can intelligently manage the home using smart plugs to turn on the washing

Sonnen may be the first company in the world to make a battery (the black object on the wall) a desirable piece of household furniture.

machine and other appliances when the PV is generating power and the battery is full of electricity. And when the domestic jobs are finished but the sun is still shining the PV is redirected to heating hot water. This is a highly sophisticated system that does everything it can to maximise the power that is generated for storage or productive use in the house. Carefully managed, Mathias told me, this can mean that 80 per cent of a home's PV production is used and only 20 per cent is exported back into the local grid.

The Sonnen battery system can manage up to 10,000 cycles of charging up and then discharging. This means that if it charges up every day and then discharges at night it will last for more than twenty-five years – almost as long as the PV on the house roof. This is far longer than people thought possible, even a few months ago.

It is little surprise that solar installers are now actively selling combined PV/battery packages, particularly in places where retail electricity prices are expensive. The cost of a 2-kilowatt hour system is now about £3,500 in Germany, which at £1,750 per kilowatt hour the system is far more costly than the figures we have seen for large car batteries, which are now down to prices of a little more than a tenth of this. However, the unit has great flexibility. Perhaps as importantly as the cost per kilowatt hour, the unit can power a building when the local electricity supply is lost. This is a very infrequent problem in Germany, where the grid remains exceedingly reliable even as renewable electricity ramps up, but is increasingly a concern in the US and in places such as South Africa where blackouts are common. The Sonnen system is modular, meaning that up to eight units can be chained together to form a storage system for a small business. Given that modern business cannot function without electricity, some companies are willing to pay high prices for the guarantee provided by the system that power will always be available.

The other value of a domestic battery system is that it can divert electricity into storage when the national grid of a country is under stress from having more solar or wind energy available than can be productively used. We can certainly expect more and more grid operators to require PV farms and smaller domestic arrays to limit their flow out into the wider grid. Batteries can help deal with this problem. Sonnen has a weather forecasting system incorporated in its software and it can predict when the house should be storing electricity and when it should be using it, for example, to run the dishwasher.

Nevertheless at £3,500 for a 2 kilowatt hour system the Sonnen battery system is not an obviously attractive financial deal. If the kit lasts twenty-five years, the cost is £140 a year, before considering maintenance costs. A unit of this size can store a maximum of 730 kilowatt hours a year (365 x 2kWh), implying a cost of about 20 pence for each kilowatt hour stored, and then used, in a UK home. And that is before thinking about the interest on the money used to buy the batteries. In other countries with higher electricity charges, the economics are better and Sonnen claims that the cost of much larger battery systems can be as low as €0.17 per kilowatt hour in Germany, well below the domestic cost of electricity. As you might expect, it is expanding into other high electricity cost locations, such as California, where householders pay increasing prices as they use more electricity each month, and Hawaii, which has the highest electricity costs in the US.

Sonnen's latest move is more radical yet. It is becoming an electricity utility. As its sales have expanded in Germany it has set up a new subsidiary in order that the owners of its battery packs can buy and sell renewable energy to each other through a central marketplace. When I have too much electricity and you have too little, Sonnen's platform sells you my excess. If it has been a cloudy December day all across Germany, the company buys in wind power or electricity from anaerobic digestion plants to fill the gap. All this is handled automatically via its central software platform and the smart meters that tell it second by second how much power your PV system is generating and how much is actually being used in the home. The promise to members of the scheme is that the electricity bought in by the utility from other generators will cost about €0.23 a kilowatt hour, a discount of about 25 per cent on current German prices (but still somewhat more expensive that fossil fuel-generated electricity in the

UK). And, in addition, Sonnen will discount the price of your battery by about 20 per cent if you join in the community of energy buyers/sellers at the same time as putting your storage system on the wall of your living room. There are many analysts who see the Sonnen power company as the prototype of how the electricity supply system will work in a couple of decades' time.

Storage of electricity at a larger scale

The people who keep very remote mobile phone base stations operating know more about the need for electricity storage than almost anybody else. In difficult terrain and often inhospitable climates, base stations have to work twenty-four hours a day. Until recently, this meant using diesel generators to provide the electricity needed. Now, this task can often be done using PV and batteries.

Pete Bishop founded PowerOasis nearly ten years ago to try to make sure that the mobile phone networks had the electricity supply to reach the billion or so potential customers living in areas with no conventional electricity supply. His business ensures that the base stations for mobile phones have independent electricity supplies to use, even in the remotest locations. That, I presume, is why the business is called the lovely name that it is. But, almost by accident, his company now seems to have moved into markets that are even more central to the world's future.

I go to see him on a grey, wet November day at the company's offices outside Swindon. I'm there to talk about batteries. But he starts by enthusiastically showing me pictures of the huge drones that Facebook plans to put into the stratosphere around the tropical world, with PowerOasis providing the software. These planes have the wingspan of a

Facebook drones, powered by PV, will provide mobile and internet coverage to areas off the regular networks.

commercial airliner but weigh a lot less than a car. The power to keep the electric engines working comes from PV panels that cover their carbon fibre wings.

Facebook's idea is to put hundreds of these pilotless planes 60,000 feet or more above regions with poor or non-existent internet coverage, well above the flight paths of commercial airliners. Don't think that this just means Africa and parts of Asia. A British mobile phone operator had been to see him in the previous week to ask about using the drone to cover the less populated parts of Scotland.

The idea is that during the day the sun charges the drone's batteries and keeps the planes circling in a continuous two-kilometre loop. Data from the ground is bounced by the drones back to remote villages where it can be collected and then redistributed to users via WiFi. The plan is to keep these huge planes in the air for three months or more. Apparently, Facebook envisages having 10,000 pilotless drones airborne at any one time. PowerOasis is handling the control systems and software for the project. (If you're worrying about whether Scotland is too gloomy to charge the batteries, you don't need to be concerned; the drones fly far above the troublesome clouds.) This is a perfect example of how PV and batteries are going to work together to do things that even five years ago we thought were impossible. Renewable energy is going to give people benefits that fossil fuels find it difficult to deliver.

I start by asking Pete about the cost of the batteries in the drones. 'Will they be cheap enough to make universal access possible?' I anxiously ask, knowing how absurdly expensive satellite-delivered internet is in remote regions. Like so many people I have talked to in the research for this book, he uses his hands to answer my question before even opening his mouth. His hand moves across his body from left to right, sinking as it goes. He's mimicking the shape of the experience curve. 'It doesn't really matter if it is not inexpensive enough today. We know it will be soon. In any event, it gives people internet at a cost an order of magnitude less than satellite.'

We go into the laboratory to look at some of the kit that will go in the drones. An ultra-light container surrounded by carbon fibre will hold tubes of Panasonic's standard lithium ion batteries, of exactly the same type as keep Tesla cars moving. The core of the container is made of a relatively inexpensive mixture of the new wonder material graphene and a wax that melts, and so absorbs energy when the cells get hot. This keeps the overall temperature down during the day and stops the cells malfunctioning. The wax then 'refreezes' at night to recreate a solid.

This is all utterly fascinating stuff but it's just an introduction to what I really want to ask about. Is lithium ion going to be the technology of choice for large scale fast-discharge applications? PowerOasis has just become Panasonic's partner for grid-level batteries around the world holding large amounts of spare power. Panasonic makes the batteries themselves and PowerOasis will provide the control systems. Despite their reputation for rugged performance, lithium ion batteries are temperamental and they need careful software control to ensure they continue to perform properly.

How are large banks of these batteries going to help maintain electricity supply when the sun isn't in the sky? We go into the next room and he shows me a rack of Panasonic battery modules sitting in a frame. It looks very much like the way computer servers fit into the racks in huge data centres. Each frame holds about 75 kilowatt hours of electricity. Ten of these merged together will form an electricity storage unit for PowerOasis's first large scale grid storage project. Unsurprisingly, it's going to be at Panasonic's new headquarters in Denver. Two standard twenty-foot containers will hold the batteries, each totalling 0.75 megawatt hours, and all the associated electronics. (Batteries are direct current (DC) devices, so like PV systems they need inverters to enable them to operate on conventional AC grids.)

These batteries will have two separate functions. The local electricity grid is paying part of the costs of the installation and in return it gets to use the battery's capacity to help smooth the voltage on the distribution network. It needs the ultra-rapid response provided by a lithium ion system to ensure that variations in renewable electricity supply, caused by an unexpected sharp drop in wind, for example, don't cause the voltage on the local network to move outside safe boundaries.

The second role of the battery system is to provide a reserve power supply to the Panasonic data centre in the event of loss of supply. If the local grid goes down, the Panasonic building will be automatically electrically isolated from the wider network, all the less important uses of electricity in the building will be disconnected and the battery power will go into ensuring that the servers and other critical pieces of kit are kept powered. PowerOasis is also providing software to manage electricity use in the building, ensuring that appliances that don't need to work all the time are turned on only when power is abundant.

In thirty years' time, my guess is that this arrangement will be typical. Every major commercial building – and perhaps most houses – will have their own battery system and will be ready to go independent of the local network for several hours or even days. Separately, the grid operator will have the right to remotely command each battery to send power in or out of the system to help keep the power supply stable. I'm unsure quite how these two occasionally conflicting needs will be managed but small scale things like this tend to be well handled by markets. (For the avoidance of doubt, I have no confidence whatsoever that the much larger process of moving the entire world on to renewables will be easily done using free markets. I think massive governmental intervention is wholly and unavoidably necessary to maximise the speed and minimise the cost.)

As I always tend to do, I ask about the price of the Denver batteries. Pete tells me that the target for this first installation is around $300 (£200) per kilowatt hour of storage capacity or about $450,000 (£300,000) for the entire 1.5 megawatt hour system when it is finished. Because of the size of the installation, that figure is a lot cheaper than Sonnen's smaller system for domestic houses. If there were no other costs, and there was no energy loss in the storage and discharge process, over 6,000 cycles the £200 of battery cost would be equivalent to just over 3 pence per kilowatt hour.

This is not trivial, but nor is it overwhelmingly expensive compared to the costs of electricity. And the costs are going to keep coming down as the car manufacturers (or 24M) scale up. In the US, experts say that at around \$150 per kilowatt hour batteries become cheaper than building new power stations to meet the daily peaks of electricity demand. In this new environment, the battery is charged up when power is readily available, and thus probably inexpensive, and is then waiting to meet the late afternoon or early evening peak.

Storage in car batteries

Earlier in this chapter I referred to a 2014 battery cost estimate by two Swedish academic authors. In the paper which presented these figures, they estimated world production of lithium ion batteries for cars had totalled less than 20 gigawatt hours of storage capacity by that year (equivalent to little over half an hour of average UK electricity consumption).

As the number of electric cars grows, they will become the biggest potential storage facility for electricity in the world. In 2016, some projections suggest that well over a million pure electric and plug-in hybrid cars will be sold. That's about 1 per cent of total global car sales. Across all types of today's electric cars, including hybrid petrol cars with small batteries as well as fully electric vehicles, the average amount of storage may be about 10 kilowatt hours, implying a total storage capacity of about 6 gigawatt hours, or less than 1 per cent of the UK's daily electricity demand.

Things will change. As batteries become cheaper, motorists will be more likely to buy all-electric, rather than hybrid, vehicles. And each car will have a larger battery pack, with a longer driving range. If we assume that by 2035 fully electric cars are 50 per cent of the total number of the vehicles on the road, then their automotive batteries will be able to provide most of the overnight storage needed by the electricity system. Perhaps you think that this is too rapid a transition to be believable but the UK, for example, would then have about fifteen million electric cars, each averaging about 50 kilowatt hours of storage and parked about 90 per cent of the time, ready to provide or to accept electricity.

Those cars will have just under a day's total electricity requirement for the entire nation stored in their batteries. If travel patterns remain the same, with cars averaging about 10,000 miles a year, then on average these 15 million cars will only actually use about one seventh of their battery capacity during the average day, meaning the rest of their power is available for use as a buffer for the grid – or to provide the same functions as a Sonnen battery by storing power from PV systems for overnight use in the house.

At the moment very few electric cars can be used as back-up batteries for the grid. Perhaps surprisingly, some are to be found in the Dutch city of Utrecht which is experimenting with using car- and bus-charging points to stabilise the local electricity system by enabling two-way flows of power. But most cars around the world are not equipped even to stop charging at times of grid stress, let alone pump electricity back into the network. This will change. Electric cars will grow in sophistication as well as in the amount of storage capacity they have, and will increasingly have the capacity to send power 'vehicle to grid' to help stabilise power networks second by second. Or even to make their owners some money by selling their electricity at times of high prices, much like the cus-tomers of OhmConnect in California get paid for turning appliances off when power is in short supply.

Most estimates suggested that 2015 installations of large-scale stationary batteries, directly attached to the grid in a similar way to PowerOasis's installation in Denver, amounted to about 500 megawatts around the world. The figure for megawatt hours wasn't available but I guess it would be about 1,000 megawatt hours or one gigawatt hour. This is less than 20 per cent of the 7 gigawatt hours of batteries contained in cars expected to be sold in 2016. And no one seems to doubt that electric cars will become more and more numerous. The case in favour of using car batteries as one of the major sources of short-term storage capacity is overwhelmingly strong. Fleets of fully electric cars with the capacity to both take electricity from the grid and return it when it is needed will be a vital part of the transition to a solar-dominated future around the world.

More reasons why battery storage will take off

It is not just that batteries are getting cheaper. Less well understood is that the widespread use of storage will make energy more expensive for those that don't have batteries. The logic for this is simple: a large part of the cost of running an electricity utility lies in ensuring that you can supply enough power at the point of peak demand on your network. You have to have enough transmission equipment as well as the surplus electricity generating plant ready to start operating, perhaps for only a few minutes, to meet maximum electricity use. Quite logically therefore commercial users in many parts of the world get charged both for the amount of electricity that they use and a second fee that is calculated by reference to their peak demand over a particular period. This is their contribution to the cost that the network bears for ensuring it will always be able to fulfil demand.

Commercial users that are subject to these 'demand charges' will cleverly use their batteries to reduce their maximum use of electricity. The business will ensure that the battery is always fully charged, either from the sun or indeed from conventional electricity sources, at the point of maximum demand.

As the peak minutes approach the battery begins to discharge to reduce the power demand in the building. So, for example, an office might find that its point of highest electricity use is at 4pm in the afternoon when the air conditioning is working at its hardest. Therefore the batteries might charge up during the day and discharge from 3.30pm to 4.30pm.

Researchers at the GTM research house quoted one executive currently installing batteries for commercial users on the impact that storage can have: 'Green Charge CEO Vic Shao said the startup's batteries can typically shave ten to fifteen per cent off of a customer's demand charges, by injecting stored power to reduce building's peaks in energy use.'

The battery owner saves money in this way. But the utility will still have to cover the full costs of the transmission infrastructure necessary to meet the highest levels of demand. So those on the network without the ability to use battery storage to hold down peak demand will end up having to pay more, particularly in countries such as the US, where pricing is regulated by government agencies. The lead author of the Moody's credit agency 2015 report on batteries said:

> Peak shaving will lower power bills for commercial and industrial customers, which will lead to regulated utilities shifting costs from battery customers to non-battery customers to recoup the revenue losses. The clear implication is that as storage becomes in wider and wider use the laggards have to pay more for their power, increasing the pressure on them to install their own batteries.

Once again, we can see that the growth of solar energy and associated storage tends to increase the financial pressure on producers and consumers of fossil fuel power.

Long-term targets

Battery technologies are improving everywhere but the US is spending most on fundamental research. The Argonne National Laboratory outside Chicago is headquarters to a key science hub trying to improve battery performance and cost. As Professor George Crabtree, the director of the programme, said, in a recent lecture, there is no shortage of approaches to try. Hundreds of varying combinations of electrolyte (the material in which the electric charge moves) and different anodes and cathodes are available. Each one has to be researched and either pushed out off the priority list or taken to the next stage of investigation.

The targets of Professor Crabtree's group are wonderfully clear. He wants to be able to hand over to his commercial partners well-defined technologies that can deliver storage at less than $100 per kilowatt hour and that can do this with a density greater than 0.4 kilowatt hours per kilogramme and per litre. His team is focusing on targets that, as he said, are 'beyond lithium ion'. The laboratory is looking for advances in battery chemistry not just improvements in manufacturing.

I spoke to him about this at the end of the lecture, suggesting that while it certainly looks as if other battery chemistries could offer better performance in terms of energy density than lithium ion, the sheer manufacturing scale of lithium ion factories means that the cost of new chemistries will struggle. Professor Crabtree didn't disagree – there is little that scientific researchers can do in a laboratory that can undercut a product that is benefiting from such a rapid learning curve in factories around the world. We may find that lithium ion – despite concerns over the ultimate energy density that is achievable, the number of charge/discharge cycles it can cope with and the very occasional fires batteries suffer – will dominate for decades to come.

When we looked at PV, we found the same phenomenon. Although other semiconductors might conceivably be better in terms of efficiency or potential future cost, the sheer momentum of silicon makes it near-impossible for other approaches, such as the organic PV of Heliatek, to keep up. The market doesn't always choose the very best option when one competitor has got so far ahead that its challengers look expensive or under-resourced.

The similarities with silicon-based PV continue. Standard single layer silicon panels won't improve much further in terms of their efficiency in converting light energy to electricity and lithium ion is also already close to its likely maximum in terms of 'energy density', the amount of electricity that can be stored per kilogramme of material. This is now around 220 watt hours per kilogramme and is highly unlikely to improve by more than 30 per cent, however much research money is thrown at the problem. A well-designed car needs about 1 kilowatt hour for each four miles driven, so 200 miles of driving range needs batteries that will weigh as much as 200 kilogrammes.

How much lithium is there?

Although many alternative battery technologies are jostling for our attention, lithium ion is probably increasing its lead over the competition, as least in the fast discharge arena. The rapid rate of cost reduction means that this battery chemistry is beginning to look unstoppable, at least for another decade or so. Professor Crabtree's target of $100 per kilowatt hour may actually be achieved by Tesla or other battery producers by the end of the decade using standard lithium ioncell construction.

Many people therefore ask the question: is there enough lithium to cover the world's needs? The answer is probably 'yes', provided that we recycle batteries at the end of their life.

The United States Geological Service (USGS) publishes estimates of the total amount of recoverable lithium around the world. The best sources are in dried up lakes in South America and, in total, USGS thinks we have at least 13.5 million tonnes of lithium to exploit. A battery containing storage of one kilowatt hour's worth of electricity needs about 150 grams of lithium metal. The arithmetic is simple: the world has enough of the metal to make batteries holding about 90 terawatt hours of electricity, or about three hours of average world use in a couple of decades. (You'll remember the estimate that global energy use will be about 30 terawatts by 2035 or so.)

This is a huge amount of battery storage. If each kilowatt hour costs $100 to make and install, the total cost will be about $9 trillion, or about 12 per cent of today's world annual income. Spread over twenty years, this is a manageable expenditure, but it will have to be paid for alongside PV and other storage media, of course.

Lithium ion batteries do not last for ever. Whether it is after a hundred, a thousand or ten thousand cycles, the batteries degrade and need to be replaced. For the world to have enough short-term storage, the lithium in them must be recovered and reused every time the battery is taken out of use. Almost no lithium is recycled today but the experts I spoke to said no technical reason blocks a move to near 100 per cent recycling in the future. As a metal – just like steel or aluminium – lithium will not be degraded during a well-functioning recycling process.

Alternatives to lithium ion

Rather than just assume that conventional lithium will dominate, at least in smaller scale fast discharge applications, I looked around for other battery types that might be viable challengers. A variant called 'lithium sulphur' is a favourite contender so I went to see David Ainsworth, the chief technology officer of Oxis, a young battery company just outside Oxford that is completely focused on this chemistry. A couple of months after my interview there, I saw the Oxis name on a list of the most innovative global battery suppliers on a desk in a car business I was visiting. This is a highly regarded company in the industry.

David started by saying that all the many different types of battery necessarily involve compromises between a large number of different characteristics such as weight, volume, safety, the voltage provided, recharging times and cost. In addition, manufacturers need to keep in mind how fast the battery loses its charge when not being used and how many times it can be successively charged and discharged before it begins to lose its capacity to accept energy. There is no single right answer to the question 'Which type of battery is best?'

He told me that lithium sulphur batteries sit at 'a position of high energy to weight, lowish cost per kilowatt hour, poorish energy to volume and a good capacity for thousands of charge/discharge cycles'. A large photograph on the wall of the offices showed a nail hammered through one of Oxis's cells so I didn't have to ask about safety. (Although lithium ion's record is good, you really wouldn't want to pierce a standard laptop battery with a six inch nail and be in the room at the same time.) The benign characteristics of lithium sulphur mean its early markets are going to be electric bikes (as the batteries are safe and light) and aerospace applications (for the same reasons). The businesses it cannot easily enter include consumer electronics and cars (because the batteries have a low energy density per unit volume and low voltage).

In the lab, Oxis is achieving about 425 watt hours per kilogramme with lithium sulphur – about twice what Panasonic manages with the conventional lithium ion batteries. But the battery wears out after a few charges and discharges. David told me that the theoretical maximum is over six times this level of energy density but this maximum will take a long time to achieve. His five-year target is to get to 500 watt hours per kilogramme and an ability to successfully charge up the battery at least 2,000 times.

So, I said, when you achieve this, what will the cost be? David replied with what he called a 'loose and wild' estimate of about $250 per kilowatt hour. If Oxis achieves this level it will be a remarkable achievement for a small venture-funded company. But think back to what GM said that it was paying for lithium ion cells – $145 per kilowatt hour – in October 2015. Including the packaging and wiring, the full battery packs today are well below Oxis's target for 2020. So lithium sulphur is going to struggle to capture the mass market for batteries able to deliver instantaneous power. There will be niches – possibly large – in which it can successfully compete, but its batteries may never be as cheap as the standardised lithium ion cells that power electric cars, phones and laptops.

Just before going out to see David, I'd been to a lecture given by Peter Bruce, one of the world's foremost academic specialists on batteries. Professor Bruce works on another type of battery, this time using lithium and the oxygen of atmospheric air. At the end of his talk he put a slide on the screen showing prospective costs and energy densities. 'Lithium air,' as it is called, promises to deliver as much as a full kilowatt hour per kilogramme of weight, perhaps three times what standard lithium ion will ever achieve. But on the screen next to this impressive number was an estimate of the target cost. At about $180 per kilowatt hour (which can only be achieved by mass manufacturing), it will still only match what is being achieved today by car manufacturers buying lithium ion cells. In other words, despite its manifestly superior qualities (even David Ainsworth, who backs the alternative lithium sulphur, describes it as the 'Holy Grail' of energy storage), it may never beat the existing standardised fast discharge lithium ion cells on price.

Large scale grid storage: PV plus battery

Primus Power, one of the leaders in large scale storage on the electricity grid, claims:

> Energy storage devices can act like a warehouse for the grid, allowing us to manage the expected ebb and flow of demand throughout the day. They can also act like shock absorbers, making intermittency of either supply or demand easier to handle.

We will need more and more of these useful attributes.

The islands of Hawaii are the test-bed for new techniques for marrying renewable generation and storage capacity. Households currently pay very high electricity charges for power that is mostly provided by expensive diesel generators. Prices are almost twice the levels of the UK. As solar PV becomes more and more important, with some parts of the islands struggling to absorb the impact of rooftop installation rates of over 20 per cent of all homes, large batteries on the grid have become urgently needed. Storage is no longer an option but a necessity.

The utility on the small island of Kaua'i announced in September 2015 that it had agreed a deal with California's SolarCity to install a 13 megawatt/52 megawatt hours lithium ion battery park to help store excess PV for the late afternoon and evening period of peak demand. This is probably the first PV plus battery installation in the world to provide what is called 'dispatchable' power. That is, when the utility needs power, it instructs the battery owners to provide it as required. This makes PV virtually equivalent to conventional fossil fuel generation. And, in this case, the battery owners are being paid less for their electricity than it would cost today for the utility to get it from conventional generators. 'PV plus battery' will therefore save householders money on their electricity bills. It will also stabilise a grid that sometimes sees three quarters of its demand met by PV and suffers occasionally from unexpected banks of cloud that can rapidly swing over the small number of large solar farms on the island. 'We have been investigating energy storage options for more than two years, and price has always been the biggest challenge,' local utility CEO David Bissell said in a statement. 'This is a breakthrough project on technology and on price that enables us to move solar energy to the peak demand hours in the evening and reduce the amount of fossil fuel we're using.'

You may say that Hawaii is almost unique. It is reasonably prosperous and yet isolated from large-scale grid power and highly reliant on expensive diesel generation. Although the islands may become the first places in the developed world to move to entirely solar PV electricity generation, they might seem to have little relevance to the rest of the globe.

That's not the case, however. Hawaii is actually very like almost everywhere in the world that doesn't today have an electricity supply or where the grid is weak or overloaded. Villages, towns, cities and regions without electricity will follow this lead and never put large-scale fossil fuel power plants in place. Instead they will watch Kaua'i and other isolated places experiment with PV and batteries and copy what these regions successfully learn.

At the other side of the American continent, PJM, the huge regional grid controlling electricity distribution from the Atlantic to the Great Lakes, is also installing batteries as fast as it can at locations right across its region. There are batteries at disused coal power stations and in scientific research facilities. Their purpose is very different to the batteries in Kaua'i. There's no immediate prospect of this network being overwhelmed by solar's daily cycle but the network still needs batteries to help match supply and demand on its grid second by second. Batteries have become cheaper (and quicker) at doing this than what used to happen – large power plants adjusting their electricity output.

PJM's projects are the beginning of very rapid expansion in the US large scale storage sector. The leading research group in the field, GTM, guesses at a more than twenty-fold expansion in the amount of storage installed in the US between 2015 and 2020. California – where legislation is driving a sharp expansion in battery use – has a 4-gigawatt battery pipeline (that's enough to provide about 10 per cent of average UK electricity demand).

Flow batteries

Most media attention is devoted to lithium ion as the energy storage device of the future but it looks to me as though the really big projects to store large amounts of energy overnight may eventually be handled by batteries using different chemistries. Some may be 'flow batteries' in which electrons flow through an otherwise impervious membrane separating two tanks of circulating liquids. These batteries have long lives and require little maintenance, although the control systems can be quite complex. Others may use a different chemical combination called 'zinc–air'. This is still virtually unknown technology but – having looked at the various businesses offering flow and zinc–air batteries – my hunch is that two stand out as potential global success stories.

Imergy, a Californian company, sells flow batteries using recycled vanadium from steelworks waste, substantially cutting the raw materials cost of its product. Vanadium is a key ingredient because in its oxidised form it has four different chemical states, more than any other chemical element. The battery works by using electrons to shift vanadium between these different forms of oxidised state.

The company has deals with electricity infrastructure suppliers around the developing world to install hundreds of batteries to power micro-grids and mobile telecommunications towers. These flow batteries can be easily scaled up in size, have low maintenance requirements and can provide decades of daily use – something lithium ion will struggle to match. They are safe, work well in high ambient temperatures and are essentially indestructible. They therefore fit well with the needs of many countries creating an electricity network for the first time in remote locations. And it has other useful attributes. Imergy has just signed a contract to provide grid batteries to Ontario, where its flow batteries will help to stabilise voltage rather than provide cover.

Imergy is coy about its prices. It writes of being able at some stage to offer prices as low as $300 per kilowatt hour but doesn't seem to be quite there yet. This makes it more expensive that lithium ion, although flow batteries will last longer. At the target price, assuming zero interest or maintenance charges, Imergy customers need a payment of just over 4 US cents per kilowatt hour over a twenty year life if the battery charges and discharges every day.

We shouldn't minimise the importance of this number. It is almost as large as the price of generating the electricity in the first place in a solar PV dominated system in a very sunny area. On the other hand, is it not an overwhelmingly large cost. Imagine a village in India with a PV system and a vanadium flow battery. The battery might have to store half of the total electricity produced, therefore adding an average of 2 US cents a kilowatt hour to the price of electricity across the entire day. This is not an insignificant increment to the price of power but means that a PV generating system, even with full overnight storage, is going to be far, far cheaper than an electricity supply based on diesel or any other non-connected supply of power other than PV.

Eos is the other battery company that I predict might be making thousands of megawatt hours of grid storage in five years' time. It makes a zinc battery that it claims is cheaper than any other type of electricity storage at around $160 per

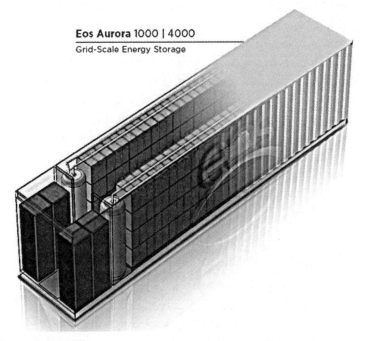

Eos Aurora 1000 | 4000
Grid-Scale Energy Storage

An EOS battery, housed in a container (it's remarkable how this sixty-year old invention is helping renewable energy due to its usefulness in protecting batteries and control equipment).

kilowatt hour. But commercial sales have not started yet, so we cannot know if this number is realistic. If it is, then Eos will capture much of the waiting market for electricity storage in places such as California and Hawaii. Safe, with high energy density, a potential life of many thousands of cycles and with rapid response and using earth-abundant materials, Eos shows how batteries will be able to provide bulk storage for electricity when it is abundant.

I've written almost exclusively about batteries in Germany and the US in this chapter. However, it would be wrong to think that grid storage was only making rapid progress in these two countries. By the end of 2015, South Korea, one of the energy 'islands' for which batteries are crucial – because it has no interconnections with any adjacent electricity grids – had committed to install 200 megawatts of storage and plausibly aims to reach 2 gigawatts of storage by 2020. A local company successfully put the largest battery farm in the world in place in early 2016 as a way to give the electricity utility more control over how it handles peak demand.

Instead of investing in more fossil fuel plants to provide grid stability as renewables expand, power utilities are increasingly using batteries to fill this role around the world. This early market for large scale static batteries is providing further impetus for steep cost reductions and manufacturing expansion.

Pumped hydro

Batteries look as though they will be by far the most important form of day-to-night electricity storage, but various other technologies also compete for this important prize. The best established by far is known as 'pumped hydro' and probably constitutes more than 95 per cent of all the large scale storage existing today around the world.

When electricity is in good supply, the operator takes water from a reservoir at a low level and pumps it uphill to a higher reservoir. At times when electricity is needed, the reverse flow occurs. The water flowing downhill passes through turbines that generate electricity.

The largest pumped hydro plant in the UK is Dinorwig, in the mountains of Snowdonia, Wales. Its reservoirs can store about 10 gigawatt hours of power and can discharge at a rate that produces about 1.7 gigawatts – very roughly, 3 per cent of UK electricity for about four and a half hours. This is a highly efficient system with a 'round trip efficiency' – the electricity generated as a percentage of the electricity used to initially pump the water uphill – of about 75 per cent. And it can usually react quite quickly – in ten seconds or so – to the changing circumstances of the UK grid, for example when a power station elsewhere unexpectedly fails.

It is in many ways a perfect form of storage. Reliability is high and pumped hydro can work daily for many decades. The problem is that it is expensive to construct and the number of sites available is limited. One developer is currently trying to finance another, much smaller, pumped hydro plant close to Dinorwig. For £160 million or so, Quarry Battery says it will create about 600 megawatt hours of storage, delivered at a rate of 100 megawatts as water is allowed to flow between two disused slate quarries. That equates to around £270/$400 per kilowatt hour, or somewhat more than the current prices for lithium ion batteries. It suffers from other disadvantages, too; its round trip efficiency will inevitably be lower, its response time to calls for power is seconds rather than the milliseconds batteries can achieve and it will cost much more to maintain. But it will last a lot longer than lithium ion cells.

Quarry Battery indicates that it believes that the UK has the capacity for at least 10 gigawatts of storage in pumped hydro locations, providing five hours of continuous power. This would offer enough electricity to meet the UK's needs for about ninety minutes and the company quotes experts saying that the country needs at least this storage capacity. This portfolio would cost, Quarry Battery says, about £16 billion. The 10 gigawatts of new reservoirs would usually be in remote places (because they need large height distances and thus tend to be in inaccessible mountains). They would also require huge numbers of new electricity transmission lines, often across beautiful landscapes.

By contrast, batteries can be placed very near to stress points on grids. In Cornwall, for example, the further expansion of renewable energies, such as wind and solar, is being blocked by a shortage of capacity on the transmission line linking the county to the rest of the UK. Batteries located at the main junction points on the grid would accept surplus Cornish solar power during the day and then discharge it at night. This could hugely increase the maximum electricity generating capability of the county.

Batteries also have a small footprint. A modern flow battery can offer 1 megawatt hour of storage or more inside a standard forty-foot shipping container. Quarry Battery's plan for 50 gigawatt hours of storage (50,000 megawatt hours) of pumped hydro storage in Welsh and Scottish mountains could be achieved instead by 50,000 containers sitting next to electricity substations, solar parks and big wind farms. Placed three-high, all of these batteries could conceivably be fitted on a total area of half a square kilometre. Or, more likely at 100 locations around the country, each around 70 by 70 metres. With today's communications capabilities, they could all be remotely told when to charge and discharge. There's no obstacle to having thousands of storage points, all sitting next to the critical substations at the edge of the grid where they add most value.

My best estimate is that the UK needs at least ten times the amount of short-term energy storage Quarry Battery suggests is necessary. That means many more batteries but, even still, the total space needed would be no more than five square kilometres. Some individual golf courses are almost as large. Of course, containers aren't pretty but shielded by trees they would be almost invisible. I admire the ambition of Quarry Battery, but the rapidly improving economics of more conventional batteries seem likely to strand their plans.

Compressed and liquid air storage

Are there other ways of storing and releasing energy quickly and efficiently that might be financially competitive with batteries? And which could work at a global scale? Two options have attracted attention in recent months. First the US company Lightsail has begun to persuasively advocate compressed air, which means that air is compressed and stored when electricity is plentiful and then released through turbines at times of shortage. In the past, scientists have been highly sceptical about this idea because when compression pressure is applied to air, it heats up. Much of the energy used in the compression is therefore lost as heat and cannot be recaptured. (Those of us who have burnt our hands on bicycle pumps understand this phenomenon well.)

But Lightsail claims to have got round this problem by injecting a very fine spray of water as the air is compressed. The water absorbs the heat and is taken away and stored (or used to provide warmth for a building, for example). When the time comes to expand the air again to generate electricity, the hot water is sprayed back into the expansion chamber, warming the air. The value of this is that warming the air adds to its pressure, meaning the turbine will rotate faster and generate more power. In addition, this is a process that is economical with regards to energy; up to 90 per cent of the electricity going in is regained in the discharge phase, a figure close to the best batteries.

The Lightsail approach is simple, efficient and extremely elegant. These characteristics will have been helpful in appealing to its roster of prestige investors including Bill Gates and Peter Thiel, the PayPal founder whose quote on the innate pessimism of most cultures ended the introduction to this book. In the spirit of Thiel's comment about the importance of optimism, I'll just say that the financial attractiveness of Lightsail's approach remains still to be seen, but I can see that the underlying simplicity of its technology could mean a very low cost in future years.

Importantly, its approach claims an energy density (in terms of megawatt hours of storage per cubic metre of space) that may match batteries. It's too early to tell, but the Lightsail compressed air route is well worth watching. Even if the company fails, the wider approach will eventually work and may be financially competitive with flow batteries. The case for further investment in this area is overwhelmingly strong.

I'm less certain about the other main contender for short-term but large-scale storage: liquid air. The central idea is similar to Lightsail's approach in using surplus electricity to compress air and then refrigerate it. The world leader, the UK's Highview Power, explains it this way:

> Air turns to liquid when refrigerated to -196°C, and can be stored in standard insulated, but unpressurised, vessels at very large scale. Exposure to ambient temperatures causes rapid re-gasification and a 700-fold expansion in volume, which is used to drive a turbine and create electricity.

This process uses standard equipment from the gas manufacturing industry, or as the company puts it, 'mature' components. The technical risks are therefore low at the first commercial scale plant currently being built in Manchester but the scope for cost reduction is inevitably limited by the fact that the equipment has already been fully developed.

So once again, the main area of concern is whether this technology can ever be truly competitive with batteries. On Highview Power's website there is a cost calculator that allows one to work out the cost per kilowatt hour of storage using its installations. For the largest plant that the web page accepts, Highview suggests a cost per kilowatt hour of around $180, or perhaps slightly better than today's lithium ion batteries for cars and equivalent to the cheapest flow batteries. But this attractively low cost level will only be achieved, the company warns, after ten installations of this size have been built. So the price may be decades away, and costs of batteries will have moved down again by then. In addition, the claimed round-trip efficiency – the amount that comes out of a storage plant

How the Manchester Highview Power liquid air plant will look.

compared to what went in – at about 60 per cent, is far lower than the yields from a comparable battery.

I was giving a presentation to a large US company in late 2015 and put Highview's diagram of the configuration for a 5-megawatt/15-megawatt hour plant up on the screen. The reaction from the engineers in the audience was quick, loud and decisive. A facility of that scale, they said, with that much steel, pipework and control engineering could never cost less than a battery bank with similar energy storage characteristics. They know a lot more than me and I suspect that they are right.

Many other types of short term storage are under investigation around the world. Flywheels, large scale compressed air reservoirs in disused salt caverns, supercapacitors and other technologies compete for the attention of investors and customers. They all have a tough road in front of them as rechargeable batteries of the lithium ion variety are sliding down an experience curve that makes them cheaper every month.

Chapter 7
Storing energy as gas or liquids

Even with millions of batteries located in homes and at electricity substations across the world, we will still need further huge quantities of energy storage. In northern Europe, for example, the hours of sunshine in the winter will provide too little power to meet electricity needs, even after energy efficiency has been improved dramatically and demand for power shifted to periods and places of abundance. PV can be strongly complemented by wind and to a certain extent by biomass, but even with this help we'll face weeks and sometimes months when electricity isn't available in sufficient quantity.

But how big a problem is this – and how much of the world's population will be able to reliably get the bulk of its power direct from the sun all year round? When we pose this question we are really asking 'what percentage of us live close enough to the equator to get enough daylight hours and reliable sun each day throughout the year?' The roughly six per cent of us who live in the UK or further north are probably too far away. There's less than seven and a half hours of daylight in the winter and the amount of solar radiation varies by a factor of almost five between the months of June and December, even in southern England. This will mean that Britain and countries to its north will need to rely on storage of energy from summer to winter.

But most of the world's population lives where sunshine varies much less during the course of the year. More than six billion – 80 per cent – of the seven and a half billion on the planet inhabit areas of sufficient daylight throughout the year: very roughly, everybody living south of New York City (700 miles nearer the equator than London). The shortest day in New York is over nine hours long and electricity generation from solar power in winter is almost half the summer level. This means that storage needs are typically far less. There are some exceptions to this rule; parts of China have relatively little sunshine, for example. Nevertheless, a surprisingly large fraction of the world's population will be able to manage well with only PV and batteries.

What about the places like the UK and other northern countries? Will a solar-based energy system really be possible? Of all the challenges facing THE SWITCH, this is the most intractable. The lack of winter sun, and the unreliability of wind, means that many people believe that PV will never provide a large part of our needs. I believe the pessimists are mistaken. The outline of the eventual solution seems already clear. At times of seasonal surplus, electricity will be used to make hydrogen using the well-understood electrolysis process. Separately, carbon dioxide will be collected, either from breakdown of plant matter or, in much larger quantities, directly from the air. Both the hydrogen and CO_2 will be fed to specially selected microbes as 'food' for their growth in large scale manufacturing plants. These microbes exude energy-rich molecules containing hydrogen, carbon and oxygen. These fuels can be kept in the existing oil and gas storage infrastructure and used when required.

Carbon neutral methane

We know that many different types of single-cell organisms naturally make methane and more complex energy-carrying molecules in their conventional metabolic processes. The task is to make these natural processes work at industrial scale and reasonable cost. The fuels collected from the workings of the metabolism of microbes will be carbon neutral, combustible energy sources that can provide power when sun or wind are scarce. They will also give us the liquid or gaseous energy sources for tasks that cannot easily be switched from conventional fossil fuels to electricity, such as aviation.

This sounds surprisingly easy and predictable, and in some ways it is. The chemistry is simple, both for making hydrogen (H_2) and for collecting CO_2 from the air. There is no shortage of candidate microbes for the generation of fuels from hydrogen and carbon dioxide. The world has not yet commercialised the production of large amounts of fuel from biological sources but several highly innovative companies are already engineering microbes that are highly efficient at absorbing H_2 and CO_2 for their growth and reproduction and then excrete energy-rich molecules that can be collected and burnt as fuel. Professor Rick Kohn, one of the leading academic researchers in this field who has looked at many different routes for using living organisms to make energy gases and liquids, told me in an email:

> Our work shows that microbes can make higher fuels from H_2 and CO_2 under the right conditions. We demonstrate that it is feasible to make a high enough concentration to make it worthwhile to extract the fuels from the fermentation broth.

The work of these new companies will mean that we will be able to store months of energy in the form of gases and liquids ready for those periods when we don't have electricity. But it may be some years before they, or their successors, are making noteworthy volumes of renewable hydrocarbons or useful alcohols. Nonetheless, some of the well-funded businesses mentioned later in this chapter will have established full-scale working plants by the end of 2016.

All these innovative newcomers face formidable challenges. The processes which they are working on are often quite inefficient. That is, the energy value of the fuels produced will seem low compared to the wattage of electricity used to generate the hydrogen and collect the CO_2 that form the raw materials. It may be an uphill struggle but I'll try to persuade you over the course of this chapter that this isn't necessarily a serious problem.

The new businesses are also impeded because the utterly vital importance of what they do is not yet fully understood by governments, regulators and investors. But that last group – investors in renewable energy – do now seem to be getting the point. Bill Gates and other wealthy individuals set up the Breakthrough Energy Coalition before the Paris climate talks in late 2015. Breakthrough will invest in a wide range of businesses focused on the development of renewable fuels, indirectly or directly created by the action of the sun.

Gates understands the crucial importance of 'solar fuels', as this class of renewable liquids and gases is increasingly called. According to the *Financial Times* he believes 'this technology could have a big impact on the global energy system because it would overcome the storage and intermittency problems of existing solar energy generators'. Gates also told the *FT* of the need to spread the investments around, 'If you fund thirty companies with a ten per cent chance of success, your chance of having a couple of successes is very, very high.' As he suggests, the critical need for renewable fuels for storage is clear but we cannot yet know exactly which approach will turn out to be the best. The good thing is that there are already many contenders trying to make solar fuels economic and large scale. I pressed Professor Rick Kohn for his guess about the eventual winner. Which microbes would turn out to be best for eating hydrogen and carbon dioxide and exuding fuels as their waste matter? His response was:

Of the six genera (types of microorganism) we isolated that make hydrocarbons, Enterococcus is very easy to handle. E. coli is often used for genetic engineering so there are methods available to easily modify it. Actinomyces is a good plant fibre digester (e.g. cellulose) if you want to use that for a carbon source. There are probably others. I don't know which I'd use just yet.

The rest of this chapter looks at some of the contenders. But first we need to take a quick look at how much long-term storage northern countries are likely to need.

Why long-term storage is needed

The chart overpage is a specific illustration from the autumn of 2015 in the UK to show why we need seasonal storage in countries with highly variable daily light levels, even if we complement PV with wind. I've plotted the running surplus and deficit in UK electricity supply from the beginning of October to the middle of November. (I picked this seven-week slot because it included two periods of unusually low wind and solar availability.) I've then modelled the actual electricity demand at each half hour point and compared this number to what would have been delivered by wind and PV if we had hugely expanded these resources at some previous time. Specifically, in this experiment, I assume we multiply

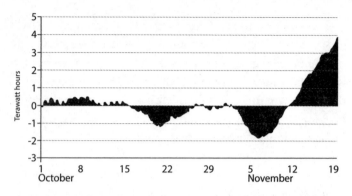

Even with huge increases in renewable generation, the UK power system will swing between surplus and deficit ever few days in autumn.

our wind power seven-fold and our PV twenty-fold.* This is aggressive expansion but well within the capacity of the UK to absorb. With this huge growth of renewables supply, how well would the UK have coped over this period?

Everything looks good for the first two weeks. Batteries and demand response would easily have covered the minor daily swings when renewables supply was mismatched with demand. Then, as the PV faded when autumn arrived, we would have hoped to see wind power expand to complement it. This didn't happen and in five days from 16–20 October even the massive expansion of renewables would not have covered total demand. In my simple model of the renewables future the UK has a maximum cumulative deficit of about 1 terawatt hour

* I have also included the actual flows of electricity from France and the Netherlands to the UK over these weeks and the UK's hydro-electric resources and burning of wood pellets for making electricity. I need to stress, too, that in this and the German charts, the demand figures are for electricity only, and not all energy, and that although we can expect further improvements in efficiency in future years we'll also see a major expansion in electricity demand to power cars and to provide heating and cooling.

– about a day's electricity need – before supply catches up with demand as the wind blows harder. Things are then okay for ten days as the deficit is recovered and as wind output fairly closely matches overall demand.

On 1 November, the wind drops again and for four and a half days the UK is severely short of power. At the nadir, the country has a cumulative deficit of almost 2 terawatt hours. Then the first of an almost interminable set of Atlantic storms arrived and electricity supply would have greatly exceeded demand. Within a week, the country would have acquired a surplus of 4 terawatt hours, over 1 per cent of its annual need.

The crucial point is this: on three occasions in a single seven-week period, batteries and other tools for matching electricity supply and demand could not have met the deficits or held the surplus. Even when every car in the UK is electric, perhaps each with as much as 50 kilowatt hours of storage, the second of the two low-wind periods would have completely exhausted all their batteries. And without large scale energy storage, we would have wasted much of the power of the ten large scale winter storms in 2015–16.

I've chosen quite a striking example here to show the scale of the problem that renewables need to address. In the UK, still periods in autumn are not unusual and the country will need many terawatt hours of storage if it is give near 100 per cent reliability to its energy supply. We can also use German figures to show the size of the storage problem. In the next chart (over-page,) I have plotted PV and wind production in Germany in each of the twelve months to October 2015. Above these two bars is the electricity generation from conventional large-scale power plants, excluding hydro-electric and production from anaerobic digestion. You'll see that although Germany obtains far more of its power from renewables than most other countries, wind and solar represented less than a quarter of its total need.

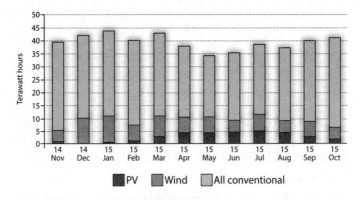

Current German electricity supply over a year.

In January 2015, for example, wind and solar combined generated about 11 terawatt hours out of a total electricity generation of 44 terawatt hours. But in October 2015 the position was even less satisfactory with wind and solar producing about 6 terawatt hours out of the total production of around 41 terawatt hours, or only about 15 per cent. (It's noteworthy that both the UK and Germany would have found it difficult to meet energy needs at the same time in October 2015 – so just adding more cables linking the main European electricity markets won't avoid the need for massive energy storage.)

My next step was to ask what combination of increased solar and wind would best fit the actual demand for electricity in Germany. The aim is to produce enough renewable energy from these sources to completely avoid using conventional power from coal and other sources over the course of a year.

One way of achieving this is to expand solar eight-fold and wind three-fold. Germany has currently more than three times as much wind power as the UK and four times as much solar, so these multiples are less than in the UK. If Germany installed these increments to its existing renewables fleet and

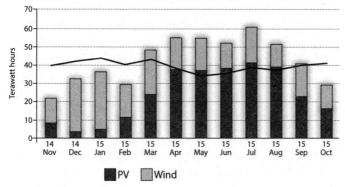

How the pattern will look as Germany continues to expand solar and wind power. The line shows typical monthly demand for electricity.

the productivity of the new assets was the same as the existing stock, then the country would see the pattern above. I've used the figures for twelve months in 2014–15.

You'll see that if Germany had had eight times as much PV in the summer of 2015 then solar alone would have met the monthly demand – give or take a few gigawatt hours – across the whole of the five-month period from April to August. Come winter, however, and even an eight-fold increase in PV only meets about 10 per cent of total power need. Even though total demand over the full year is met by the expansion of PV and wind, the months of October to February are not covered. In the case of November 2014, a month of unusually low wind would have left Germany with only about half of its electricity needs supplied. No amount of energy efficiency or demand response will stop this being a serious problem.

Over the course of the deficit months, the Germans would have to meet about 58 terawatt hours of power need from the surpluses generated in the summer months. It will need a storage mechanism that can hold energy from sometime in September to about the beginning of March in a typical year.

In theory, that energy could be held in batteries sitting in containers scattered around Germany: 58 terawatt hours would mean 58 million shipping containers each containing 1 megawatt hour. But that's twice the number of containers currently moving freight around the world. Not only would this be expensive but it would also be grossly inefficient. Each battery would be filled up once during the summer and then discharged at some point over the winter. (In addition, we'd need all the batteries for overnight storage as well.)

We have to find another method of storage or, perhaps, invest in so much PV and wind that they would cover the needs of Germany in a month like November 2014 when its power requirements were 20 terawatt hours greater than the availability from low carbon sources. According to my calculations, to meet all of November 2014's requirements would have meant having eight times as much wind capacity as Germany has today. Combined with eight times as much PV, this would, over the full year, have given the country about twice as much power as it needs.

This may be the right thing to do, but covering the winter months by hugely investing in wind will approximately double the cost to Germany of the remainder of its energy transition. What might we do instead to provide stored power?

The four main storage options

The best way of providing storage of months' worth of energy is either to use plants and trees as the source of seasonal storage or to convert electricity into hydrogen in periods of substantial surplus, possibly then using the gas to make storable fuels, either methane or liquids. Four possible options are:

a Transform plant matter into fuels to provide a buffer of energy in the form of liquid fuels.

b When power is in surplus, generate hydrogen using electrolysis and store it for seasonal use.

c Employ micro-organisms to convert hydrogen from electrolysis to methane, which is then stored in natural gas networks.

d Use microbes to take streams of CO_2 and hydrogen from electrolysis to create liquid hydrocarbons and burnable alcohols.

The central problems with these options are that in the case of *a* and *c* the process uses large quantities of biomass, and the world probably cannot hope to grow anywhere near enough plants and trees to deal with longer term storage needs (although Mike Mason's plans for shorter term storage from plants grown on otherwise useless land are plausible). Option *b* is difficult because hydrogen storage is complex and expensive. Options *c* and *d* suffer because they require sources of concentrated CO_2. Although streams of carbon dioxide are easily found today from fossil fuel combustion, they will be less readily available in a world of entirely renewable energy.

The right way forward, it seems to me, is to find a means of capturing CO_2 directly from the air and use this to generate methane, the main constituent of natural gas, under option *c* and also liquid fuels under option *d*.

Using biomass to store liquid hydrocarbons

A seasonal buffer of energy can be created by converting plants and trees into liquids similar to petrol, which can then be placed into storage and burnt as needed.

The 'biofuels' (liquid hydrocarbons and alcohols made from biological sources) that the world is currently producing are largely made from food. During growth the plant carrying the food has captured CO_2 via the photosynthesis process and converted it to sugars and carbohydrates. These compounds have a high energy content, which is why they have value to us as food. In many places around the world the simple sugars in corn and wheat are fermented into fuels that can substitute for petrol. These are usually called 'first generation' biofuels because they have used these easily fermentable foods to make liquids. When combusted, the CO_2 initially captured by the plant is returned to the atmosphere. So, in theory, using foods for making fuel can be done at near carbon neutrality.

But using food for storing chemical energy in transport fuels is not a good means to decarbonise the world economy. The average person eats about 2.5 kilowatt hours of food energy each day, while a car driver in Western Europe uses about ten times that amount of energy in the form of motor fuel alone in the same period. So even if all the food in the world was fermented into a fuel, it would not come close to providing the energy needed to keep internal combustion engines on the road. And using foodstuffs for energy also tends to increase the price of food for people whose survival is already precarious.

What are the biological alternatives to using foods as the raw material for our seasonal storage of energy? As discussed earlier, in theory the world might have enough other biological material (biomass) to make it the source for some of our long-term stores. The world's population currently needs less than 1 terawatt of the energy in food out of the approximately 90 terawatts of energy transferred each year to plants and trees via photosynthesis. Another 2 or 3 terawatts is used to feed meat animals. Some of the rest of the energy stored in plants and trees could give us what we need.

At the moment, we can convert the simple molecules in foods reasonably efficiently into liquid biofuels. But breaking down the more complex non-food molecules of cellulose and lignin in plants is far more difficult than turning sugars in grains into ethanol. The enzymes that are used to break down these much more intractable molecules are very expensive. So progress in making fuel from cellulose has been slow, and financially painful, particularly so since low oil prices have depressed the value of these petrol substitutes.

Sponsored by the national government, refineries in Brazil have gone the furthest towards commercial development of fuels made from cellulose rather than food. These are often called 'second generation' biofuels. Two large facilities run by GranBio and Raizen, a company partly owned by Shell, now successfully produce tens of millions of litres a year of fuel ethanol made from the cellulose in sugar cane stalks. That's still insignificant in world terms; it would have to be tens of billions of litres to represent a measurable percentage of total fuel consumption for cars.

Rather than using enzymes to crack cellulose and lignin into smaller and more manageable molecules the American company Red Rock Biofuels, based in Colorado, uses an alternative technology to produce liquid fuels. During 2016 it will build a refinery which takes forest waste, including pine needles and small branches, from around Lakeview in Oregon and turns it into diesel and jet fuel. It will do this by heating the wastes in the absence of oxygen to many hundreds of degrees. The woody material completely breaks down and turns into gases – largely hydrogen and carbon monoxide – in a mixture usually called 'syngas', similar to the gases produced by the Entrade gasifier (see p. 145).

The syngas can then be recombined into more complicated hydrocarbons called alkanes which are similar to the main constituents of diesel fuel. This is done using the Fischer-Tropsch process, a well-known and mature piece of chemical engineering that was invented in Germany almost a hundred years ago. Fischer-Tropsch refineries were used in South

Africa to make fuels from coal during the apartheid era when it was expensive or impossible for the country to import oil.

As the Red Rock founders are at great pains to point out, there is nothing particularly technically sophisticated in what they are doing and their fuels should be competitive even with cheaper oil. Their experience lies not in the sophisticated chemistry of biofuels but in the construction of big and reliable refineries completed to tight budgets. In this case, the cost of a refinery producing 60 million litres of fuel a year will be about $200 million. (As an indication of the scale of this plant, 60 million litres would be enough to meet the needs of about 60,000 European cars.) This is a relatively low risk process and I'd be surprised if Red Rock Biofuels isn't producing substantial quantities of diesel-equivalent and jet fuels there by the end of 2016. It already has two contracts to supply a low-cost US airline and the freight distribution company FedEx that will consume its entire yearly production.

The refinery was sited to gain access to the substantial acreages of private forest around Lakeview. The trees, mostly Ponderosa pine and white fir, are sustainably harvested under FSC (Forest Stewardship Council) rules, meaning that new growth will match or exceed the harvest each year. Red Rock will use the lower grade forest products, such as sawdust, pine needles and small branches that otherwise have little commercial value. The environmental credentials of a process using just the waste products from forests are good.

Most commentators assume that the fuels produced from these waste materials will be used for air and land transport. But a fully decarbonised world will probably mean that most surface transport will be running on electric motors, not internal combustion engines. Even ignoring the CO_2 emissions and urban air quality benefits, we want to switch transport to electricity because it is approximately three times as energy efficient. A petrol car might possibly convert 25 per cent of the energy in petrol into motion, an electric car can manage up to 90 per cent. It may be a means, however, of making airline fuel carbon neutral. Whereas we can certainly switch all surface transport to using electricity, and possibly liquefied natural gas in heavy vehicles, it will take many years before we can hope to replace liquid fuels in aeroplanes. Although air travel is growing in importance, it still only represents about 6 per cent of the total demand for oil. This means that we may well be able to make enough airline fuel from biological sources to completely replace conventional oil.

But rather than seeing Red Rock as producer of diesel substitutes, could its technology help store energy from one season to another in countries with unreliable sun? The answer is probably not, due to its potential scale of operation. The refinery Red Rock is building in Lakeview will produce fuel with an approximate energy value of about 0.6 terawatt hours a year. If combusted in large power stations at times of electricity shortage, these oils would produce power of about half this value. The seasonal electricity deficit of around 58 terawatt hours that I suggested would remain after a massive further expansion of German renewables would therefore require almost two hundred Red Rock refineries of the size of the Lakeview prototype.

Another approach to making liquid hydrocarbons from waste biomass has been pioneered by Cool Planet, a company founded in Colorado. Like Red Rock, it heats up raw materials to very high temperatures in the absence of oxygen. This process – pyrolysis – drives off gas which can then be condensed to form high energy value liquids that resemble petrol or diesel. What is left after all the volatile gases have been driven off is 'biochar', a form of almost pure carbon which has enormous value in helping tropical soils retain fertility and moisture. Cool Planet, which has a range of big investors including Google, oil companies and electric utilities, will open its first commercial scale manufacturing plant in 2017 in Louisiana. Like Red Rock, it will utilise otherwise valueless wood thinnings from commercial forests, so it can also reasonably claim that its processes do not involve the need to cut down more trees.

The US has large resources of forested land that can provide otherwise unusable wastes for biomass refining. And in large areas like northern Russia, which have extreme variations in solar availability from season to season, trees are often abundant and could be gasified to offer long-term storage in liquid hydrocarbons. Over half Russia's massive land area is forested and very little of this is exploited today. About 900 million hectares is wooded and the amount of biomass used for energy conversion could be expanded many times without reducing the carbon captured in trees. Similarly almost 35 per cent of the US land mass is under forest. Intelligent and careful exploitation for making renewable liquids makes good sense.

However, many crowded industrial countries don't have easy access to more plant matter. A large fraction of the land in the UK, for example, is already used for agriculture of one form or another with about 25 per cent of the country given over to arable crops and another 50 per cent to pastureland, principally for cattle and sheep.

However, even in the crowded UK, it should be possible to produce some renewable fuels by converting low grade pasture back to woodland. Pasture is often an extremely unproductive form of agriculture in terms of the food value gained from each hectare of land. The uplands of Wales, for example, sustain relatively small numbers of sheep across large tracts of land on which little else grows other than a thin grass. Without those sheep, as environmental writer George Monbiot has made us aware in his influential work on rewilding, woodlands would generally quietly return and prosper.

Would a sustained programme of reforestation provide the UK with enough stored energy in trees to help overcome seasonal shortages of power? The numbers aren't very encouraging. A hectare that fed a couple of sheep would grow at least two tonnes of wood a year once woodland was re-established. If I have done my numbers correctly, the useful food calories in the meat of those two sheep would be around 1 per cent of the energy we could get from the trees grown on the same area of land. Doubling the percentage of the land area of Wales devoted to trees – up from about 15 to 30 per cent – would produce wood with an energy value of around 3 terawatt hours, about 1 per cent of UK electricity need, if converted using the Red Rock or Cool Planet processes. In other words, even substantial changes to land use could not produce enough energy to cope with seasonal energy deficits in very densely populated countries like the UK. In other countries, it might make a much bigger difference.

Using electrolysis to generate hydrogen

The conversion of biomass to liquids has some value in giving less-densely populated countries large buffers of stored energy. However, it is far from a complete solution. A more feasible mechanism is through the production of hydrogen, an energy dense gas that can be stored. You simply take your spare electricity in times of surplus, and use it to electrolyse water, splitting it into oxygen and hydrogen. (You probably did something like this in the chemistry classes at school – it really is a very simple process done at a small scale.)

The hydrogen (H_2) can be stored and used either in combustion or, more probably, in fuel cells that recombine the hydrogen with oxygen and create a flow of electricity as they do. The experts have a point; storing energy as H_2 is a relatively efficient process, involving losses of as little as 20 per cent in the form of waste heat.

The problem is that it is expensive to hold hydrogen in a tank. Although it contains a lot of usable energy per unit of weight, it is poor when considered in terms of volume. Even liquefied hydrogen only has about a quarter of the energy density of the same volume of liquid hydrocarbons such as petrol. In addition, making hydrogen liquid either requires very high pressure or very low temperatures. Neither is cheap to achieve. The advance of what is optimistically called the hydrogen economy is likely also to be impeded by public concerns over the safety of large amounts of hydrogen stored, for example, as the fuel for cars.

Nevertheless, several countries are experimenting with hydrogen cars, and with hydrogen refuelling stations. The estimates I have seen suggest a current cost of about £1 million for each high pressure storage location, helping to explain why there are still only about 150 hydrogen refuelling stations around the world. Nor can hydrogen be cheaply shipped from place to place. Pipelines are at least three times as expensive as natural gas pipelines, partly through the need to protect against the long-term corrosive impact of hydrogen on steels.

Some countries allow small amounts of hydrogen to be added to natural gas networks. In Germany, for example, H_2 can be injected into the local grids until it forms a maximum of 2 per cent of the gas transported to users. Temporarily unwanted electricity is now being used to split water into hydrogen and oxygen at several sites. The hydrogen is then injected into natural gas pipelines.

Adding hydrogen to the UK gas grid is forbidden, so one of the world leaders in electrolysis for disposing of surplus electricity, ITM Power of Sheffield, is successfully focusing its efforts on selling to the power companies in Germany. It recently commissioned a 150 kilowatt plant at Ibbenbueren in the west of the country. Remarkably the Ibbenbueren site has an overall conversion efficiency of 86 per cent including heat recapture, an almost unprecedented level.

Hydrogen can also be stored in other forms, such as when it is combined into hydrides. But the amount of hydrogen contained in these chemicals is also small, expressed either in terms of volume or of weight.

The problems of energy density and possible safety issues could in theory be avoided by using hydrogen chemically combined with nitrogen in the form of ammonia (NH3). Making ammonia is well understood and because it has been the basis of artificial fertilisers using the Haber Bosch process for over a hundred years, manufacturing technologies are mature and relatively inexpensive. Unlike pure hydrogen, ammonia becomes liquid at relatively low pressures and can be used in conventional internal combustion engines, usually with a small amount of carbon-based fuel for better burning. Some commercial vehicles in Belgium during the Second World War used ammonia as fuel after diesel became unavailable, reportedly without any accidents, even though ammonia is highly toxic.

Possibly we should be more interested in using ammonia in power stations. Ammonia will burn in gas turbines, but research has been limited and the long-term viability of this route is not yet conclusively proven. If science can show how surplus energy can be used to electrolyse hydrogen, which is then converted into ammonia, stored and then burnt in a turbine the world has a viable zero carbon technology for producing electricity on demand.

This would be most useful if ammonia could be manufactured in times of surplus PV electricity and then used when the winter deficit arrives. How much ammonia could be stored for seasonal use? Would it be able to cover the 58 terawatt hour winter deficit in Germany, for example, that we saw in the calculations above? To give a sense of scale, Germany has the capacity to store about 700 terawatt hours of oil. (The UK's oil holding capacity is probably little more than 5 per cent of this because we have used North Sea oilfields as our 'reservoirs'.) Germany's storage is provided partly by tanks above ground and also by substantial use of underground caverns. The energy density of ammonia is only about half that of oil so 58 terawatt hours would need the space equivalent of nearly 120 terawatt hours of oil. Very roughly, storing enough ammonia to cover German winter electricity deficits would only use about 20 per cent of the capacity required for today's oil storage.

Unfortunately, it gets more complicated. The ammonia would need to be pressurised – albeit only to a modest level – to remain as a liquid and therefore existing large oil tanks would be of no use. To make ammonia useful as an energy storage medium would require a large new infrastructure. This would be far cheaper than a network of liquid hydrogen storage tanks, pipelines and refuelling points, but still highly expensive.

As a result, some commentators who have looked at the problems are now highly sceptical about whether hydrogen (either in its molecular form or combined with nitrogen in ammonia) can provide the cheap seasonal storage that some of the world needs to meet energy deficits when the sun is low in the sky. The 'hydrogen economy' about which many people have enthused as a way of completely avoiding the use of carbon-based fuels looks difficult and expensive to achieve. Former US Energy Secretary Steven Chu famously noted that cost-effective hydrogen fuel cells (which convert hydrogen back into water and generate an electric current in the process) required four miracles to happen. 'If you need four miracles, that's unlikely: saints only need three miracles,' he concluded.

Converting the hydrogen to methane

If hydrogen isn't going to be the medium through which we store power for months on end, what might actually work? We need to find ways of converting unwanted electricity into energy-rich hydrocarbon gases or liquid fuels that we can actually hold for months in existing storage facilities for use when needed.

One possible solution to the long-term storage problem is to combine hydrogen with carbon dioxide to make methane in a reaction that is known as 'power to gas', or increasingly just as 'P2G'. This transformation is a simple and well-understood chemical process, called the Sabatier reaction after the French chemist who first described it a century ago:

$CO_2 + 4H_2 > CH_4 + 2H_2O$
Carbon dioxide + hydrogen makes methane + water

Methane is by far the largest constituent of natural gas and so P2G will produce an energy carrier that can be immediately injected into conventional gas pipelines. In many industrial countries the gas grid already has the capacity to accept the enormous amounts of methane that could flow from a large and persistent surplus of PV-generated electricity. The network of pipelines holds many months' worth of gas consumption, ready to be used to meet seasonal needs for electricity or heating. When Germany, for example, moves into November and the five month period of insufficient power from solar PV starts, it can turn on conventional gas power stations and convert that methane back into electricity.

The German gas grid has storage capacity equal to about 200 days' average use while France has a network equivalent to about 120 days' consumption. More importantly, both countries have good capacity to supply gas in very cold winter weather when demand is much higher. In France's case this is over sixty days while Germany can manage for even longer. Apart from crude oil and refined oil products, for which many countries also have substantial storage capacity, natural gas is the only energy carrier for which there is an existing large-scale infrastructure. That is why I think this route for storage of massive amounts of seasonal energy surplus makes far better sense than using spare electricity to make pure hydrogen. Indeed, I think 'power to gas' is a technology that the world urgently needs to make work on a large scale if it is to manage full decarbonisation of energy.

The first successful plant to use surplus electricity to create methane for the gas grid has been operating in the town of Werlte in Lower Saxony since 2013. It is operated by the car manufacturer Audi and is used to provide a renewable fuel for the models that use compressed natural gas in their fuel tanks. A 6-megawatt electrolysis facility makes hydrogen when electricity is abundant but stops when power is scarce. This flexibility means it is also being paid to help grid stability as a part of Germany's 'demand response' capability.

Where does the CO_2 come from for methane production? Biogas coming from a neighbouring anaerobic digester containing about 60 per cent methane and 40 per cent carbon dioxide is piped in. The hydrogen from the electrolysis is combined in the Sabatier process with the CO_2 in the biogas to produce a gas upgraded to 100 per cent methane. Surplus heat from the plant is used for a local heating network. Because the CO_2 comes from sources that have naturally absorbed the gas in the photosynthesis process, it is nearly carbon neutral.

Other countries are also beginning to realise the importance of 'power to gas'. In the US, the first electricity to methane trial happened during 2015, its project partners including the southern Californian gas utility SoCalGas and two government agencies. SoCalGas summarised the main advantages of P2G quite neatly in its press release:

> Power-to-gas offers longer term storage capacity, cost-effectively using existing natural gas infrastructure to potentially create the world's largest storage technology. In addition, power-to-gas storage can conserve the significant amount of energy currently wasted when renewable production exceeds consumption.

P2G pioneers: Electrochaea

Elsewhere, several start-up businesses are working on commercialisation of P2G and one young German company seems to me to be the business to watch. At a waste water treatment plant outside Copenhagen, the highly innovative Electrochaea is installing its first commercially sized facility to electrolyse water to produce hydrogen and combine it with carbon dioxide to produce methane in a process that uses microbes rather than the conventional Sabatier reaction.

I spoke to project manager Dominic Hofstetter to find out how the new installation was going and understand better how the technology works. He ran an earlier trial installation in northern Denmark before moving on to this vital commercial pilot and told me that, like anaerobic digesters, waste water plants produce a sewage gas which is part methane and part carbon dioxide. Usually called 'biogas', this isn't pure enough to be put into the gas grid and so it has to be combusted inefficiently on-site in a turbine, generating electricity.

'Our aim is simple,' said Dominic. 'We want all the CO_2 from biological processes to be converted to 100 per cent methane so that it can be directly injected into gas pipelines.'

Denmark has frequent winter days on which its total electricity demand is met by wind turbines. It is aware that the surrounding countries, which currently accept its cheap exported power during gales, will eventually also have surpluses at the same time. Atlantic storms increase wind output across the whole of northern Europe. There is usually a few hours' difference between the period of peak wind speeds on the west coast of Ireland and when the gales hit Denmark. But, generally, when it is windy in Denmark is it also blowing hard across the British Isles, northern Germany, the Netherlands, Norway and Sweden. Large scale energy storage is becoming increasingly urgent if Denmark is to avoid having to frequently disconnect its wind farms to avoid having too much unwanted power on its grid. Batteries will be of limited help in capturing the excess electricity resulting from week-long storms.

This is where Electrochaea comes in. Up to 1 megawatt of electricity is used to split hydrogen and oxygen at the Copenhagen waste water plant. The oxygen is usefully recycled into the water treatment works, where it helps increase the speed of the chemical breakdown of sewage. Electrochaea then takes the pure hydrogen and passes it through what Dominic calls a 'reactor', along with biogas containing methane and unwanted CO_2. In the reactor are tiny single-celled microorganisms called archaea which thrive in conditions without oxygen. The microbes, which have been selectively bred but not genetically modified, rapidly 'eat' the hydrogen and the carbon dioxide in the two streams of gas entering their home and exude methane. The gas leaving the chamber is almost pure methane and can be immediately added to the local gas distribution pipeline.

This chemical reaction that produces the methane is useful to the micro-organism because it provides the energy to drive its metabolism. In plants and trees, this energy is provided by the sun, which drives the photosynthetic reactions. Archaea usually live in places where the sun doesn't shine, such as the human gut, and combining CO_2 and hydrogen provides the alternative source of energy they need.

Dominic and I talked about the finances of the new Copenhagen plant. The 1 megawatt electrolyser from the Belgium firm Hydrogenics cost about €1.4 million (slightly more than a PV farm of similar size would cost). On top of this, the reactor cost about €200,000, while associated pipework and control systems added substantially to the bill. Including construction, the estimated total cost was about €4.5 million.

Dominic commented that today the world market for commercially sized electrolysis plants is tiny. Only about 30 megawatts are sold each year, mostly for the production of hydrogen for the chemicals industry. In his view, we can expect huge reductions in cost as the industry grows in scale over the next few years and electrolysis undergoes a sharp experience curve. An engineer from Siemens said the same: they expect the manufacturing of the crucial membranes for electrolysis to move within a couple of years from being expensive and time-consuming to being highly automated and quick. The cost will come down rapidly and the eventual target price for a full-scale 5-megawatt plant is expected to be about €1.5 million per megawatt of electricity input, or about one third of current levels. That will make it very competitive with batteries. And, because it is a process that can handle a continuous flow of hydrogen and CO_2 over days and weeks (unlike a battery which can only store a set amount of power) it will be much more useful for long term storage.

At present Electrochaea is focusing on persuading the operators of waste water and AD plants to use its technology to upgrade their biogas to 100 per cent methane. Because the gas from these sites comes from biological sources, mostly agricultural waste and human sewage, and not from fossil sources, the extra methane output can be described as very low carbon. Other sources of biogas can also be exploited, such as from waste at food factories and agricultural discards. But the amount of biogas ready to be upgraded from a mix of methane and CO_2 to pure methane is not a large fraction of total gas use. In the UK, for example, one well-informed estimate is that the total amount of energy available from the rotting of all putrescible material into methane is about 40 terawatt hours a year. That figure can be compared to annual UK electricity production of around 320 terawatt hours. It's a reasonable amount but may not be not sufficient on its own to cover winter shortages.

Dominic is used to scepticism about Electrochaea. The first problem that people always raise is that the company's solution transforms electricity, which is usually seen as an expensive power source, into gas, which is perceived as cheap. In the UK, for example, the retail price of electricity is about three times the price of gas. And the process of conversion isn't ever going to be completely efficient: energy is lost as heat both in the electrolysis stage and when the hydrogen and CO_2 is being converted to methane by the microbes. Dominic admits that the process is never likely to get much more than two thirds of the energy value of the electricity into the energy value of the upgraded methane.

For the sceptics, it gets worse. If we are using summer electricity to produce gas to burn in combined cycle gas turbine plants in the winter, then we are also losing energy in this process as well. Even the best combined cycle gas turbine plants (CCGT) only convert about 60 per cent of the energy value of methane into electricity. So, in total, we only get back about 40 per cent of the electricity we put in at the beginning of the process.

I don't think this is the right way of evaluating power to gas. Imagine that we used the very productive Electrochaea process to facilitate the storage of some of Germany's summer surplus of power to meet the 58-terawatt-hour electricity deficit in the winter. Between May and August, much daytime electricity will be essentially valueless. Even now, there's some-times far too much to be used. The same is true in the UK. When the wind was blowing earlier today, I noticed that the price of power in the UK wholesale markets fell to well below zero. That is, users were paid money to take more electricity. We are only just beginning to see the full effects of increased renewable penetration on European grids and periods of very low prices will become increasingly frequent. I believe the falling cost of solar power means it is almost inevitable that wholesale electricity will become cheaper and cheaper over coming decades. It will matter less and less that we convert power inefficiently into gases for the storage.

Electrochaea's methane manufacturing process will convert very low value summer electricity from PV into more valu-able gas for use to generate winter electricity. The plants will be performing a useful function by stabilising the energy mar-ket at both times of the year. Prices in winter will not spike as frequently and negative prices at times of high wind or full sun will also be less severe. This value is gained even if the conversion processes of power to gas result in the loss of large amounts of the energy value. In fact, it is almost a good thing that the production of methane is inefficient because it uses up more of the otherwise nearly worthless surplus electricity.

There is a secondary argument which I think deserves much more discussion. I showed earlier in the book how electricity use can be temporarily reduced by demand response technologies such as those employed by REstore (see p. 149). However, there may always be unforeseen periods of weeks, or perhaps months, when battery storage and demand response is insufficient. This can happen even in summer if the sun is clouded and wind speeds are low.

To cope with these periods countries will inevitably need to have electricity generating capacity that stands ready to meet the needs of households and businesses. I suspect that in the UK, for example, gas-fired power stations will need to be on standby all the time, even though they may actually operate less than a hundred hours a year to meet emergency needs. This makes them extraordinarily expensive to build and operate. If, in addition to this emergency working, they also operate during the times of seasonal electricity deficit using Electrochaea's low carbon gas, it becomes much cheaper to operate these plants for each kilowatt hour of output over the year. Their running costs will be spread over much larger amounts of electricity output. This adds to the powerful financial logic of encouraging 'power to gas' technologies.

So far, we've looked at Electrochaea's 'biological methanation', as it is known, purely as a technology for upgrading biogas, either from sewage plants or anaerobic digesters, to pure methane. The logic from the company's point of view is that this is the easiest early market for it to address. But this probably isn't enough to provide all the storage we need, particularly as we move to using electricity for almost all our energy needs. We need to start finding other sources of CO_2 that can be converted to methane using hydrogen from electrolysis at times of electricity surplus.

Later, I spoke to Electrochaea's chief technology officer, Doris Hafenbradl about this. She said that the company's approach would also work well at cement plants, which are

Doris Hafenbradl with the Electrochaea reactor, which in April 2016
produced its first batch of methane.

prodigious producers of CO_2 from the chemical reactions
which break down calcium carbonate. Cement manufacture
is responsible for about 5 per cent of world CO_2 emissions, so
this may become an important industry to focus upon. The
idea is that when electricity is cheap, such as at night or at
times of seasonal surplus, the carbon dioxide from cement
plants could be piped through the Electrochaea reactor for
conversion to methane. Doris thought this might mean that
the Electrochaea system would be used about 60 per cent of
the time at most of the sites they are looking to sell to. But

to make financial sense, she commented, this would probably require a full-scale carbon tax, something the world is only hesitantly moving towards. When this happens, the owners of an Electrochaea plant would be paid to absorb CO_2, because this delivers a valuable global benefit.

While we can additionally use the streams of CO_2 coming off industrial processes such as cement-making, this cannot be a long-term solution because the gas comes from burning fossil fuels. This will need to cease as soon as possible. (Ways of producing cement without producing extra carbon dioxide are reasonably well advanced.) Processes such as Electrochaea's conversion of the CO_2 from rotting biomass into methane using surplus electricity will be hugely helpful. But this cannot be the entire solution. There isn't enough biological material around the world to provide enough carbon dioxide from combustion or gasification. We will need to supplement 'power to gas' with renewable liquid fuels.

Using microbes to make liquid hydrocarbons

Liquid hydrocarbons store much more energy than batteries, judged either by volume or by weight. The US government's ambitions for battery storage (see p. 194) are to achieve 400 watt hours (0.4 kilowatt hours) per kilogramme and per litre. Oxis, the leader in lithium sulphur batteries, is targeting 550 watt hours (0.55 kilowatt hours) per litre in five years' time. By contrast, petrol, a typical liquid fuel, generates heat energy of around 10 kilowatt hours per litre, twenty times as much. And there is, of course, the convenience of carbon-based fuels, which flow easily, can be pumped from one place to another and do not deteriorate significantly. Other than their easy flammability, they are nearly perfect as a means of holding dense energy in a form for easy use.

As with natural gas, developed economies have large amounts of existing capacity to store liquid hydrocarbons, mostly in oil storage tanks and pipelines. Even the UK, which has never invested significantly in energy storage infrastructure, can hold about 13 billion litres in already existing facilities dotted around the country. They are linked by pipelines. If we can find a way of creating renewable liquid fuels, most countries already have the infrastructure ready to accept all that can be produced, and then used when needed.

Companies like Cool Planet and Red Rock (see p. 223–4) promise us supplies of energy-rich fuels made from biological material. The next phase of development of liquid fuels will be the use of bacteria or other microbes to turn streams of gases from industrial plants and power stations, usually CO_2 or carbon monoxide, into much larger molecules with value as fuels. Most approaches to achieving this use biotechnology of various types, although several very different pathways are being explored.

Probably the best advanced is the technology being pioneered by LanzaTech, a company founded in New Zealand that is now headquartered near Chicago. LanzaTech takes a stream of waste gas from an industrial process, such as a blast furnace or a cement factory, and passes it through a bioreactor containing microbes. These microbes use the carbon monoxide from the flue gas to run their metabolism, creating valuable fuels as waste products. Under certain circumstances these bugs can also utilise carbon dioxide and hydrogen. As the company says:

> LanzaTech's process involves biological conversion of carbon to products through gas fermentation. Using microbes that grow on gases (rather than sugars, as in traditional fermentation), carbon-rich waste gases and residues are transformed

into useful liquid commodities, used in everyday applications, providing a novel approach to carbon capture and reuse.

LanzaTech's activities are based around 'carbon capture' of the carbon monoxide in waste gas streams and its storage in the form of a liquid. Its choice of microbe is from the four-billion-year-old family of acetogens, one of the earliest life forms that evolved long before any forms of bacteria or algae. Acetogens use only the simple carbon gases and hydrogen for their metabolic processes.

Emissions from many industries, including steel manufacturing, are very similar to the gases produced by hydrothermal vents that acetogens grow on in nature. However, Lanzatech first located the microbe it uses not in a hot ocean vent but in the intestinal tracts of rabbits. The company describes this bug,

A LanzaTech fermentation scientist preparing samples from a steel mill fermentation broth.

Clostridium Autoethanogenum, as its 'rock star' due to its ability to use what the world doesn't want – carbon-based gases from dirty fossil fuel combustion – as its raw material for the production of fuels. Ethanol, a conventional fuel better known as alcohol, is produced naturally as the waste product of its metabolic processes. Ethanol can be easily biologically converted to several other valuable hydrocarbons, such as ingredients for nylon and other plastics and for synthetic rubber. And, of course, the LanzaTech process involves no diversion of land or other resources into the production of liquid fuels.

LanzaTech has raised over $200 million in four separate funding rounds over the ten years of its existence. Backed by industrial companies such as Siemens, as well as traditional venture capital sources from the US, China, Malaysia and Japan, the business is widely seen as one of the most bankable enterprises in the next phase of the transition away from fossil fuels. It has pilot plants in China, Taiwan and the US but its move to full scale commercial operation will happen at the Arcelor Mittal steelworks in Ghent, Belgium.

A steelworks produces a waste gas that contains large amounts of carbon monoxide. This is usually burnt in a flare that converts it to carbon dioxide. So if the carbon monoxide is used in the LanzaTech process and converted to useful ethanol and other chemicals it will reduce the net emissions of the blast furnace. At the Ghent works, the €87 million project will result in the production of about 50,000 tonnes a year of ethanol containing enough energy to fuel about 30,000 cars. Eventually scaled up across all of Arcelor Mittal's European steel operations, the impact would be ten times as big.

Other potential sources of the carbon monoxide, carbon dioxide and hydrogen that LanzaTech's bugs need include municipal waste and biomass residues. We don't know yet how much the ethanol and other products coming from the company's fermentation process will cost but the interest shown by steel companies around the world suggests that they are interested in using their waste gases to make ethanol for sound commercial reasons, not just the environmental kudos they will get from avoiding emissions.

The LanzaTech process isn't a perfect solution to the need for liquid hydrocarbons as stored energy. First, it needs a single source of large scale carbon emissions, and second it doesn't convert a net 100 per cent of those gases into a useful liquid. Some CO_2 is added to the atmosphere, too, although much less than would be the case without the fermentation process.

Another promising company, Joule Unlimited, based in Massachusetts, has the first of these problems but could, at least in theory, avoid the second. Joule uses a different microbe to turn the CO_2 in a stream of waste gas into ethanol or other liquid hydrocarbons. Their process bubbles gases containing carbon dioxide through water exposed to the sun in long tubes. (This doesn't need to be clean, fresh water, so there is no impact on water availability). Conventional cyanobacteria, often called 'blue-green algae', would use the sunlight and the CO_2 to grow and reproduce. However, Joule's engineered bacteria use the photosynthetic process to create ethanol, or several other useful liquid hydrocarbons. These hydrocarbons can be separated from water in the final phase of the process.

Like LanzaTech, Joule has already raised about $200m to take its idea to the point where it can plan a commercial plant in 2016. These are not cheap ideas to develop and it will take time – probably around ten years – to get the technologies to a stage at which full-scale production plants can be built. So it is fortunate that Bill Gates and his new group of billionaire business 'angels' are concentrating on this field.

Joule promises us extraordinary results for full-scale commercial operations on the basis of its pilot plant in New

Mexico. It claims it will be able to produce about 250,000 litres of ethanol a year from a hectare of tubes carrying water, CO_2 and its proprietary bacteria. The energy value of that amount of ethanol is about as much as the electricity that could be generated by PV installed on the same area. The cost of the fuel will be competitive with $50 oil, the company says. And its plants can be installed on arid lands that cannot be used for any other purpose. There is no impact on agriculture and its use of potable water is minimal.

It is fair to say that Joule Unlimited has as many sceptics as suppporters. Concerns include the amount of energy needed to separate the ethanol from the water, the continuous risk of invasion of algae into the system (which would destroy its productivity), and an open disbelief at the levels of energy conversion claimed to be achievable by the Joule microbes. Most living plants are unable to convert more than 1 per cent of the light energy falling on their leaves into the complex molecules they need to grow but Joule says its microbes can achieve over 10 per cent photosynthetic efficiency, a level close to the theoretical maximum.

Joule's pilot plant is in sunny New Mexico where it uses the flue gases of local fossil fuel manufacturing plants. Like Lanzatech, future commercial farms will be located next to steelworks and cement factories in order to get a large supply of CO_2 for the microbes to feed off. In another claim that its detractors find difficult to believe, the company says that a full-scale 4,000 hectare plant next to a source of CO_2 could generate a volume of liquid hydrocarbons equivalent to a 50 million barrel oil field.

Algenol is a third company using small biological organisms, again cyanobacteria, to absorb CO_2 and turn it into ethanol. As with Joule, the company's technology uses sunlight falling on transparent plastic tubes containing the water and the microbes. Its trial plant in Florida claims to deliver fuels for about 35p a litre, almost competitive with oil, even at low 2016 prices. As with Joule and LanzaTech, Algenol has taken the best part of a decade to get this far and has absorbed large amounts of its investors' money. However, the process works and many experts believe it can produce highly competitive fuels from seawater, a stream of waste CO_2, sun and obliging microbes.

As an economist by training, not a scientist, I always tend to ask about incentives. Organisms, whether microbes or people, tend to need inducements before they do something. I asked Professor Rick Kohn from the University of Maryland why the micro-organisms used by these pioneer companies were prepared to do something useful to humanity by excreting liquid fuels. In essence, my question was 'why do these microbes let us have the energy? Surely they want to keep it themselves?' Rick replied that the micro-organisms capture energy themselves from the chemical reactions inside their cells. They use the energy to grow and make proteins and other useful material. What's left is useful to us but not necessarily to the microbes. He also points to the importance of ensuring that the quantities of 'foods' of CO_2 and hydrogen available are conducive to the maximum possible production of fuels for humankind. Broadly speaking, the more CO_2 and H_2 we give them, the more fuels we will get back.

Artificial photosynthesis

In the next few years we will also see increased emphasis on directly using solar energy to get living organisms to produce useful precursors to standard liquid fuels – a process that is sometimes called 'artificial photosynthesis'. Many people talk of the need to increase investment in clean tech research and

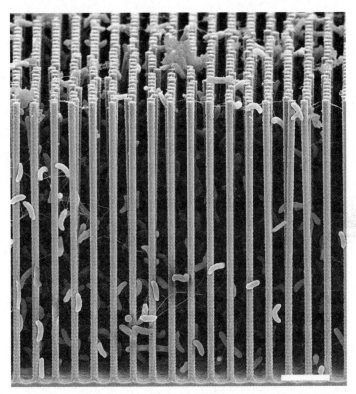

A scanning electron microscope photograph of a nanowire array populated by *Sporomusa Ovata* bacteria.

improving artificial photosynthesis should be an important destination of such cash. PV technologies already have enough momentum behind them and large amounts of risk money should instead be flooded into 'sunlight to liquids' research.

One of the latest advances has been reported by a team at University of California, Berkeley. Their research shows how light hitting tiny vertical nanowires (these are spikes only a few nanometres wide) gives electrons enough energy to free them from their usual position in an atom's orbit. A bacterium, *Sporomusa Ovata*, is one of several types of living organism that can use these electrons to capture CO_2 in the liquid solution and convert it to a slightly more complex molecule, acetate, which contains two carbon atoms. The bonding of the two carbon atoms in the acetate ion is a crucial first step towards the creation of much larger and more useful liquid fuels.

The image on the previous page is from a very powerful microscope and shows the banana-shaped bacteria gripping the rods in the forest of nanowires. Their aim in hugging the silicon and titanium oxide wires is to gain access to the useful electrons. As a side benefit, the nanowire homes also enable the bacteria to survive in a stream of flue gas that still contains some oxygen (which would rapidly kill them in normal circumstances). In the experiment that formed the basis of the research each cell could produce about a million acetate ions per second when the nanowires were under full illumination. *Sporomusa Ovata* is unusual in being focused on consuming CO_2 (and just CO_2) when it is available.

The effective conversion efficiency is still low at about 0.4 per cent of the light energy entering the nanowire forest – slightly less than a leaf typically manages. But the team at Berkeley is planning a new experiment that will raise the efficiency up to 3 per cent. This is still far below the best solar panels, which can turn 20 per cent or more of the energy of light into electricity, but nevertheless a big step towards economical production of liquid hydrocarbons and alcohols for storage purposes.

'Photosynthesis can solve the energy conversion and storage problem in one step,' asserts Professor Peidong Yang, one of the investigators working on the project. 'It converts and stores solar energy in the chemical bonds of organic molecules. Once we reach a conversion efficiency of ten per cent in a cost effective manner, the technology should be commercially viable.' As with Joule's technology, the only requirement is brackish water and trace amounts of vitamins and the production plants of these companies can be put in locations where there are few, if any, alternative uses for the land. Professor Yang believes that the system 'has the potential to fundamentally change the chemical and oil industry in that we can produce chemicals and fuels in a totally renewable way'.

Air capture of CO_2

All the processes for creating liquid and gaseous hydrocarbons discussed so far need concentrated streams of waste CO_2 from biological materials, factories or power plants to feed the microbes making liquid fuels. This is because the density of CO_2 in atmospheric air is so low. We won't get enough solar fuels if the pioneer companies do not feed the micro-organisms with the CO_2 and H_2 raw materials they need at a high enough concentration.

This is a problem. As discussed, we probably don't have enough easily available plant matter to use as a source of CO_2 coming either from pyrolysis or anaerobic digestion. And we really don't want manufacturing or power generation facilities to be using fossil fuels at all. The reason is that any use of oil, gas or coal will inevitably add to the CO_2 in the atmosphere.

Here's why. Imagine a power station from which we capture all the CO_2 from its combustion of natural gas. If we feed this carbon dioxide and some hydrogen from electrolysis to the right microbes in conditions which allow them to thrive, they will make liquid fuels for our energy storage needs. But the point of these fuels is that they will eventually be burnt, at times of energy shortage, recreating that CO_2 and putting it back in the atmosphere. To be carbon neutral, artificial pho-tosynthesis for seasonal storage needs a source of CO_2 that has not come from fossil fuels. It needs to be captured from the air around us.

CO_2 is less than one part in two thousand of air, so this task is also full of complexity and expense. Technologists have long been extremely pessimistic about direct air capture of CO_2. Plants can do it, of course, to feed their photosynthesis activities but so far humankind has not mastered the same skill. However, the last couple of years has seen increasing optimism that separating out the tiny quantities of CO_2 from the nitrogen and oxygen in ambient air will prove to be possible at a large scale and – eventually – at a reasonable cost.

Leading the field, it appears, is Climeworks, a spin-out from the Zurich Federal Institute of Technology, one of the top technical universities in the world. Climeworks uses a chemical called an amine that absorbs CO_2 as air flows over it. Amine absorption is a well-known technology already used in oil refineries. When a sufficient amount has been collected, the absorbent is heated and nearly pure CO_2 is driven off and collected. The amine that has gathered the CO_2 is then returned to the capture area and can be continuously reused.

As a result of the Climeworks process, we have a means of generating a concentrated stream of carbon dioxide that can be used to combine with hydrogen from electrolysis to make more complex carbon molecules for use as fuels. When solar PV energy is abundant, we can make hydrogen and then hydrocarbon liquids. This can either be used directly as a primary energy source – for transport purposes, for example – or can be combusted to generate electricity.

Climeworks' first commercial plant is currently being built at Hinwil, near Zurich. The installation sits on top of a waste incinerator, not because it employs the CO_2-rich exhaust gases but because the unit uses its heat during the stage of the process when CO_2 is driven off the absorber material. The installation is intended to capture about 1,000 tonnes of CO_2 a year, all of which is destined for a local horticultural greenhouse.

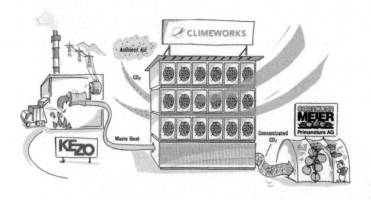

A virtuous circle: Climeworks' first air capture plant will use concentrated CO_2 in greenhouses to increase plant growth.

CO_2 pumped into a greenhouse assists plant growth by raising the rate of photosynthesis – you might get 30 per cent larger lettuces in the same temperature and light conditions by supplementing the carbon dioxide in the air.

At first sight, the economics of this process look really terrible. About €4 million was needed to design and build this CO_2 capture plant, although much of it was a one off cost because it was the first commercial installation. In return the developers get 1,000 tonnes of pure CO_2 a year, probably worth at current market prices less than €50,000. This represents a financial yield of little more than 1 per cent of the cost of the plant and omits any consideration of the energy costs of running the machine. Electricity needs are estimated at 200–300 kilowatt hours per tonne of CO_2 (perhaps costing €20 or more at today's prices). In addition, substantial amounts of heat are needed to drive off the gas after it has been absorbed, although in many industrial locations 'low temperature heat' (about 100° C) is readily available and has little other value.

At the moment the company estimates the underlying cost of the CO_2 from its capture plants is about $600 a tonne. That's between ten and fifteen times the price that carbon dioxide would sell for in a commercial market today, although there is only a limited amount of manufactured carbon dioxide actually available. Another benchmark is the price heavy industrial users and power companies have to pay for CO_2 emissions permits in the European trading scheme. This is currently about €8 a tonne, or little more than 1 per cent of the cost of CO_2 directly captured by the Climeworks machines.

I spoke to Christoph Gebald, one of the company's founders, to ask him about the costs and how these might be reduced in future projects. The scale of the challenge doesn't unduly concern the young founders of Climeworks, he said. A few years ago, the American Physical Society, a US group of physicists, published its estimates of how much direct air capture of CO_2 would cost. It saw difficulties getting the cost below $600 a tonne of CO_2 for a plant capturing a million tonnes a year. So the Climeworks machine is already producing carbon dioxide at the same price as the physicists said was only possible with a plant one thousand times as big. When Climeworks scales up, it will push the costs of CO_2 capture down to levels as low as $150 a tonne or less.

How does Christoph see the Climeworks technology being used to create energy-rich liquid fuels for energy storage? He says that it can either work with biological methods of creating fuels, including the technologies of companies like Joule or Electrochaea, or it can use more simple chemistry. An electrolyser that produces hydrogen can be used alongside its machine to induce a chemical reaction that produces 'syngas', the mixture of hydrogen and carbon monoxide. As we saw earlier (p. 223), the syngas can be put through the Fischer-Tropsch process to generate liquid hydrocarbons. To industrial chemists, this is all relatively easy. The combination of the Climeworks technique for generating CO_2 and the availability of hydrogen electrolysed by renewable electricity will mean that we can create entirely carbon neutral energy-rich liquids similar to oil.

The Harvard physicist David Keith started a company in Calgary, Alberta, to do something very similar to Climeworks. His venture, Carbon Engineering, is building its first plant to capture CO_2 by passing air through a solution containing potassium hydroxide. The liquid chemically absorbs the carbon dioxide in the air, creating potassium carbonate. Pure CO_2 is then driven off the carbonate in the second and third phases of the process, using much higher temperatures than the Climeworks approach. The chemistry is simple but the process engineering is demanding. Nevertheless, according to an interview in *Nature* magazine, David Keith estimated the costs of CO_2 capture in commercial plants could be as low as $100–200 a tonne.

But why would anyone ever decide to pay that amount of money? The reason most often proposed is that if the world ever decides to impose a tax on carbon, CO_2 capture would qualify as offsetting, or counterbalancing, other sources of emissions. So a power plant, for example, could reduce its carbon tax by drawing in CO_2 from the air to counterbalance the emissions from burning fossil fuels. However, it seems unlikely that the tax would ever be high enough to make it worthwhile to reduce a company's net liabilities.

For my money, the reason why air capture of CO_2 will eventually take off is that the world will eventually find it cheaper to make liquid fuels from CO_2 than to use oil. Christoph Gebald from Climeworks told me that a tonne of carbon dioxide will create about a third as much of a high quality liquid hydrocarbon such as petrol after being fed into a process such as Joule's, for example. This is worth about $160 at today's oil prices – a price most analysts expect to rise.

So, is CO$_2$ capture viable?

There are many 'ifs' here, but if direct CO$_2$ capture becomes very cheap and the world could manufacture solar fuels based, for example, on the Joule Unlimited technology, the inventions of Berkeley's Professor Yang or Electrochaea's methane-producing bugs, then it is perfectly possible that direct capture of CO$_2$ would become financially viable, even without a carbon tax. This would be helped considerably if the carbon capture from the air was done using very low cost electricity at times of grid surplus.

Running through this book is a consistent theme: that the conventional view of renewable energy as inherently more expensive than fossil fuels is mistaken. Energy from the sun is becoming cheaper than other sources and in a couple of decades' time it will be much cheaper still. The reason that direct capture of CO$_2$ from the air appears to be expensive today is only that the process uses large amounts of energy. When energy is a fraction of today's price, as it surely will be in decades to come, the obstacles to collecting CO$_2$ from the air and using it alongside hydrogen to make renewable fuels largely disappear. The processes for generating renewable fuels described in this chapter are fundamentally inexpensive, once input energy costs are excluded. Whether it is Electrochaea's archaea bioreactor or LanzaTech's acetogens, the biological factories for renewable fuels are relatively simple compared, for example, to an oil refinery. This is why storage in the form of renewable gases and liquids looks set to be the inevitable and financially beneficial means by which the world will deal with seasonal surpluses and deficits.

Epilogue
Some conclusions

Vaclav Smil, Bill Gates's favourite author, suggests that transitions from one fuel to another take a half century or more. The move from wood to coal took this long and the shift from coal to oil and gas about the same. Solar PV currently produces less than 1 per cent of the world's energy requirements and in Smil's view we are therefore near the beginning of at least a fifty-year transition to renewable energy. In his book *Energy Myths and Realities* (2010) he concluded that 'the inertia of existing massive and expensive energy infrastructures and prime movers and the time and capital investment needed for putting in place new convertors and new networks make it inevitable that the primary energy supply of most modern nations will contain a significant component of fossil fuels for decades to come'.

I believe Smil is wrong and that THE SWITCH will be far quicker and simpler than previous transitions. Solar PV is cheap, and becoming cheaper every day. It outcompetes other forms of energy supply in large areas of the world already. More and more types of energy use – particularly surface transport and heating – can be electrified. Unlike even a decade ago, electricity use can be turned up and down automatically in response to the availability of solar energy. The energy system is becoming more and more like a conventional market which uses price to adjust demand so that it matches supply. All of these factors makes PV's growth much easier to accommodate.

For countries with good levels of sunshine around the year, which probably house as many as six billion of the world's current seven billion population, the solar revolution may need little else to fully replace all fossil fuels. Elsewhere, in northern Europe and parts of the US, the complements to PV, including wind, CSP and some forms of biomass, can be used to provide power at night or when sunshine is in short supply. Short-term storage – lithium ion and other batteries – is falling in price almost as fast as PV and will become extremely cheap and very widely used within ten years.

Longer term storage is in an earlier phase of development. We know that hydrolysis of water to make hydrogen from surplus power is going to provide a critical component of THE SWITCH. This hydrogen, which will in effect be almost free because of the low (or even negative) wholesale price of electricity in daytime, will be used to make methane. The chemical reactions to achieve this are simple and do not require large amounts of external energy in order to proceed. All that is required is reasonably sensible government policies that encourage the development of a widespread 'power to gas' infrastructure. This gives us renewable gas for storage in the huge existing network of pipes, tanks and caverns to be used to generate electricity when the sun is shining.

We can also see the way to inexpensive use of hydrogen and carbon dioxide to make renewable liquid fuels. Several types of microbes have been identified that can make liquid fuels for us if we feed them hydrogen and CO_2 and several successful young companies possess the intellectual property to create petrol and diesel substitutes through these processes. These renewable oils can be stored in the same pipes and tanks in which refined oil products are held today for consumption. Much more work needs to be done but the fundamental chemistry is not hugely complex and many billions of dollars of capital are available for research and development. There is no reason to believe that these renewable liquids will be any more expensive than today's petrol and other oil-based fuels.

An important reason for optimism about storage is the overwhelming downward pressure on wholesale electricity prices produced by solar (and wind) abundance. We are used to thinking of energy as expensive but as solar becomes more and more omnipresent, it will drive down the cost of power everywhere. We want this. It means that the key ingredient we need in order to make renewable gases and liquids is very low cost. As a result of lowered electricity prices, natural gas and petrol substitutes will themselves become unexpectedly cheap. We know that low-cost (or even free) electricity will make hydrolysis very inexpensive. It will also hugely improve the economics of collecting carbon dioxide directly from the air.

As importantly, and this is not something extensively discussed by Smil, we will need no new physical infrastructure in the form of tanks, storage caverns and pipes for storing and moving these zero carbon fuels. Smil makes the correct assertion that energy transitions normally need new networks, meaning such things as pipelines and electric grids. One of the most positive aspects of THE SWITCH is that the cost of this will be tiny in the case of the move to PV. It will be the first energy transition that has ever occurred that does not require a new distribution network.

Equally important, for poorer countries with weak or non-existent electricity grids there will be no need to build a large infrastructure for electricity distribution. Instead entrepreneurs will construct energy 'islands' for villages and towns, based around PV farms and batteries for overnight storage. As with the arrival of the mobile phone, people will get access to a vital service without having to wait decades for government and large companies to build infrastructure.

In other words, there is no significant obstacle to THE SWITCH. It can happen in a couple of decades. Money will be saved, air pollution will be cut and the threat from climate change reduced by a transition to solar energy sources. PV will also give faster and cheaper access to electricity than any fossil fuel alternative for the unconnected 1.3 billion people on the planet and for the billion or so others who suffer from unreliable supply.

What seemed impossible to energy experts as little as a couple of years ago is now – extraordinarily – within our grasp. Societies therefore face a choice of deciding whether or not to drive towards an affordable renewable future or to recoil at the short-term difficulties of the transition and drift towards climate nightmares. It is actually no choice at all. Uncomfortable or not, the world needs to move as fast as it can towards the future described by Ben van Beurden, CEO of Shell, in which solar will be the central provider of the world's power needs.

Appendix
Figures and terms

There are a lot of numbers in this book. As the late Professor David MacKay, the UK's most influential writer on energy, said, adjectives aren't enough when discussing the energy problem. But the numbers are mainly about the relative size of the problems and solutions discussed, and there is very little that is difficult or technical. Indeed, I think the only technical things are two related concepts: the difference between power and energy, and the daily solar power curve.

The difference between power and energy

Throughout this book, I write about two different ideas: the *flow of power at a particular moment* and the *total amount of energy a machine/home/country uses over a certain time period*. The flow of power is expressed in multiples of 'kilowatts' (1,000 watts). The amount of energy is described in multiples of 'kilowatt hours' – 1 kilowatt sustained for one hour. So a machine which uses 3 kilowatts (a tumble dryer, say) consumes 3 kilowatt hours over the course of sixty minutes. For larger flows, the world uses a hierarchy of different units:

A kilowatt is 1,000 watts

A megawatt (MW) is 1,000 kilowatts

A gigawatt (GW) is 1,000 megawatts

A terawatt (TW) is 1,000 gigawatts

So a terawatt is a billion (a thousand million) kilowatts. And a terawatt hour (TWh) is a billion kilowatt hours.

Generally, talking about individuals or domestic homes, the right unit is a *kilowatt of power*. For example, over the course of the day the typical domestic home in Europe is using about a third of a kilowatt of electricity. The figure in Australia is about twice this number and in the US four times. (This isn't the total energy consumption of these homes. There may be gas or oil for heating, and the residents may drive a car.) The appropriate unit for a country is *gigawatts*. For example, as I write this on a Sunday afternoon, German electricity production is about 42 gigawatts: i.e. during this hour Germany will use 42 GW hours of electricity. If this number stayed the same over the full day, then the total electricity used would be 42 GW x 24 or about 1,000 *gigawatt hours*. That's one *terawatt hour*, the next unit up.

In the UK, the wholesale price of electricity is usually quoted in £ *per megawatt hour*. It is about £40 at present and less in the US and most of Europe. The retail price tends to be specified in *pence per kilowatt hour*. Today, a careful UK shopper can probably buy a package at around 11 pence a kilowatt hour; it is higher in the US and much lower than Germany. 11 pence per kilowatt hour is equivalent, by the way to £110 per megawatt hour. The average home in the UK uses about 3,000–4,000 kilowatt hours or 3–4 megawatt hours.

The daily solar power curve

On to the daily *solar power curve*. An array of solar panels sitting on the roof of a home might have a peak power of four kilowatts. That's the maximum flow of electricity generated when the sun is shining on to the panels at midday in high summer. The energy output, even on a cloudless day, will be less in the early morning and late afternoon when the sun is lower in the sky and therefore the sun's light energy is hitting the surface of the panel at an angle.

At night, the panels sit there idling. But during the midday hours in June and July, the production of these 4 kilowatt hours of panels is far greater than is needed by the house, which might be little more than one third of a kilowatt at that time. A house with a

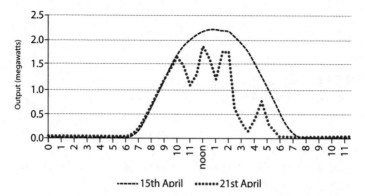

------ 15th April   •••••• 21st April

The output of electricity from a solar farm on two days in April 2014.

battery could store this excess but mostly the electricity flows out into the public grid, to be consumed by other households. The chart opposite shows the output of a 2.5 MW solar farm in the south of England on two very different days in April 2014. As you can see, its performance follows a completely smooth curve on April 15th – a constantly sunny day. The 21st of April, by contrast, started well but went downhill at around 10am; output then went up and down for a few hours until the sky turned black at around 3.30pm, when the electricity generated fell to almost nothing for an hour or so.

The other important things to note are that the length of the producing day is about seven hours in winter and seventeen hours in summer and that the peak on a completely sunny day in December is only about 60 per cent of the maximum in high summer. And, of course, the number of December days in the UK in which the sun shines continuously are few.

Over the course of a year, a solar panel in a good location in the southern part of the UK will produce about 11 per cent of its theoretical maximum (if the sun were shining right down on to it twenty-four hours a day). In other places around the world the average is probably 20 per cent or so and I use this global number at various places in the book.

Thanks

I am grateful to the following generous people who gave their time to talk to me or allowed me to test my ideas:

Kurt Adelburger, David Ainsworth, Mike Berners-Lee, Thomas Bickl, Pete Bishop, Mathias Bloch, Brenda Boardman, Chris Case, Steve Connolly, Doyne Farmer, Kevin Fiske, Christoph Gebald, Kat Glover, Nick Goddard, Philipp Grünewald, Doris Hafenbradl, Dominic Hofstetter, Nick Jelley, Nina Klein, Rick Kohn, Chris Llewellyn Smith, Owen Loudon, Mark Lynas, Mike Mason, Malcolm McCulloch, Pieter-Jan Mermans, Robin Morris, Andy Moylan, Richard Nourse, Mark O'Hare, Michael Parker, Arno Schmickler, Peter Sermol, Bill Weil, Gage Williams, Naomi Wise, Steve Zadesky.

Thanks also to the people at Profile Books who helped immeasurably with the structuring and editing of this book. Mark Ellingham was patient, constructive and always insightful. Andrew Franklin's letter confirming the commissioning of the book was an inspiration to me. Martin Lubikowski produced admirably clear charts; Carol Anderson proofread; and Caroline Wilding indexed.

I owe the greatest debt to my wife Charlotte Brewer and daughters Alice Brewer, Mimi Goodall and Ursula Brewer.

Images

Grateful acknowledgment is made to the following sources for providing images:

Page 2 (Roger Easton) NASA; p. 13 (diagram data) BP; p. 69 Chris Case; p. 72 Nina Klein; p. 81 Oxford Photovoltaics; p. 85 Heliatek; p. 87 Heliatek; p. 95 Wikimedia Commons; p. 118 Solar Reserve; p. 121 Helio100; p. 129 Spinetic; p. 136 Tropical Power; p. 140 Tropical Power; p. 169 Energiesprong; p. 175 Getty Images; p. 177 Tesla; p. 180 24M; p. 182 Sonnen; p. 186 Facebook; p. 203 Eos; p. 209 Highview Power; p. 239 Doris Hafenbradl; p. 242 LanzaTech; p. 247 University of California, Berkeley; p. 251 Climeworks.

Index

A

Abramovitz, Yosef 64
absorption chilling 146
acetogens 242–3
Actinomyces 215
Africa
  food production 134
  solar power 60–4
  see also individual countries
AGL 7
agriculture
  pumping operations 56–7
  UK 226–7
Ainsworth, David 197–8, 199
air, carbon capture 249–54
air travel 224–5
algae 244–5
Algenol 245–6
alkanes 223
aluminium 108
American Physical Society 252
amine absorption 250
ammonia 229–30
anaerobic digestion (AD) 5, 131,
  132, 135–41
Andasol 117
Andhra Pradesh 54
Apple 66, 67, 108
Arcelor Mittal 155, 243
archaea 234–5
Argonne National Laboratory 194
Arriba 154
artificial photosynthesis 246–9

asphalt 153
Atacama desert 59
Audi 232
Austin, Texas 53
Australia
  domestic electricity consumption
  260
  tracking 96

B

bacteria 247–9
Bangladesh 16–17
banks
  funding solar 105, 106–7
  interest rates 98–100
  solar power predictions 50–1
batteries 5–6, 173, 240
  car batteries 190–2
  cost declines 44, 173, 176–9, 256
  and demand charges 192–3
  domestic 57, 171, 181–5
  drones 185–7
  flow batteries 174, 201–3, 206
  grid storage 44, 187–90, 206–7,
    219–20
  lithium air 198–9
  lithium ion 173–96, 199–201, 210
  lithium sulphur 197–8
  long term targets 194–5
  PV plus battery 199–201
  and time-of-use-pricing 162
  24M 179–80
  zinc-air batteries 201, 203–4

A note on references

The vast majority of references for my research on this book are online and I have therefore provided links to these on my website, for ease of use. Please see:

www.carboncommentary.com

blood brother

Also by Michael Simmons:

pool boy